SHAPING THE FUTURE

Biology and Human Values

by Steve Olson

Board on Biology
Commission on Life Sciences
National Research Council

NATIONAL ACADEMY PRESS
Washington, D.C.   1989

NATIONAL ACADEMY PRESS ● 2101 Constitution Avenue, NW ● Washington, DC 20418

This book is based on a symposium sponsored by the National Academy of Sciences. It has been reviewed according to procedures approved by a Report Review Committee consisting of members of the National Academy of Sciences, the National Academy of Engineering, and the Institute of Medicine.

The National Academy of Sciences is a private, self-perpetuating society of distinguished scholars in scientific and engineering research, dedicated to the furtherance of science and technology and their use for the general welfare. Under the authority of its congressional charter of 1863, the Academy has a working mandate that calls upon it to advise the federal government on scientific and technical matters. Dr. Frank Press is President of the National Academy of Sciences.

The National Research Council was established by the National Academy of Sciences in 1916 to associate the broad community of science and technology with the Academy's purposes of furthering knowledge and of advising the federal government. The Council operates in accordance with general policies determined by the Academy under the authority of its congressional charter of 1863. The Council has become the principal operating agency of both the National Academy of Sciences and the National Academy of Engineering in the conduct of their services to the government, the public, and the scientific and engineering communities. It is administered jointly by both Academies and the Institute of Medicine.

Funding for the symposium and for the production of this book was provided by the National Research Council Fund, a pool of private, discretionary, nonfederal funds that is used to support a program of Academy-initiated studies of national issues in which science and technology figure significantly. The NRC Fund consists of contributions from a consortium of private foundations including the Carnegie Corporation of New York, the Charles E. Culpeper Foundation, the William and Flora Hewlett Foundation, the John D. and Catherine T. MacArthur Foundation, the Andrew W. Mellon Foundation, the Rockefeller Foundation, and the Alfred P. Sloan Foundation; the Academy Industry Program, which seeks annual contributions from companies that are concerned with the health of U.S. science and technology and with public policy issues with technological content; and the National Academy of Sciences and the National Academy of Engineering endowments.

The views in this book are those of the speakers at the symposium and the author and do not necessarily reflect the views of the National Academy of Sciences or the National Research Council.

Library of Congress Cataloging-in-Publication Data

Olson, Steve, 1956—
    Shaping the future : biology and human values / by
Steve Olson for the Board on Biology.
        p.   cm.
    Includes index.
    ISBN 0-309-03947-9.—ISBN 0-309-03944-4 (pbk.)
    1. Biology. 2. Biology—Moral and ethical aspects.   I. National
Research Council (U.S.). Board on Biology.   II. Title.
QH311.044   1989   89-12565
574—dc20   CIP

Foreword

In studying the philosophers Benedict Spinoza and Immanuel Kant, we have come to realize the extent to which their investigations of metaphysics were a prelude to their ethics. For both of these men it was necessary to know what the world of nature was "really" like before asking what the world of human relations should be.

But Spinoza on the early side and Kant on the late bracketed the Newtonian revolution. Since that time it has been impossible to form a view of what the world is like without including physics along with epistemological and ontological ideas. With the publication of Darwin's *Origin of Species*, biology was joined to those disciplines that constitute the *is* that precedes the *ought to*. At first, Social Darwinism became an explanatory theory of ethics, then an epithet, and now it returns to the arena of discourse in other forms.

We are now about a century and a half past the Darwinian revolution, and ecology, ethology, and molecular biology have become parts of that now vast domain that must be understood as part of our attempts to lead a good life and fulfill our duties to others. The query "What is life?" is very much a part of asking, "How should it be lived?"

In light of the preceding, it seems altogether fitting that one of the events marking the opening of the Arnold and Mabel Beckman Center of the National Academies of Sciences and Engineering should be a symposium on biological research and human values. Such an event provides a chance to ask what we are and then turn to the vital question of who we are. The first query is a part of science proper; the second interfaces with ethics. And just as science constantly changes with new developments, we come to the startling discovery that ethics are not rigid but must also alter in response to our changing understanding of the world and the new technological *milieu extérieur* that we continually create. The vast social and environmental consequences of scientific discovery change the intellectual setting within which the discourse takes place.

Thus, the ethical aspects of our inquiry are twofold: we still have the traditional problem of understanding the world and using that understanding to determine how we should act. We also have a second obligation imposed because our scientific understanding of the world leads, through technology, to profound alterations of that world. As we have been instructed by allegory for a long, long time, tasting of the fruit of the tree of knowledge is not necessarily an ethically neutral act.

Biologists have two tasks in regard to human values. The first is to help in the constant exploration of how ethics must respond to new views of the biological world that emerge from laboratories and field researches. The second is ethically to monitor the effects of science and technology on our present activities and the future of human society. This book has many things to say about both issues. Few questions can be fully answered, but it is important that they are confronted. Scientists are now willing to talk self-consciously about ethics. That in itself is an important development.

The present state of the planet and the rapidity of developments lend a certain urgency to scientists' concerns with ethics. Thus, this book on biology and human values comes at a propitious time. One is reminded of the words of a much earlier ethicist, Rabbi Hillel: "If not now, when?"

Harold Morowitz
George Mason University

Preface

In 1988 the National Academies of Sciences and Engineering opened a new facility in Irvine, California—the Arnold and Mabel Beckman Center—to serve as a west coast conference center for their members and study groups and for those of their associated organizations, the Institute of Medicine and the National Research Council. To commemorate the opening, a series of symposia were held on critical scientific and technical issues facing modern society. One considered the technologies and public policies affecting future sources of energy. Another looked at the state of mathematical literacy in the United States. And a third, on biological research and human values, furnished the basis for this book.

As often happens within the National Academy of Sciences complex, the theme for the symposium derived in part from an earlier project. In 1985 the Committee on Research Opportunities in Biology, a unit of the Board on Biology under the National Research Council's Commission on Life Sciences, undertook the gargantuan task of surveying all of biology, from molecular genetics to ecology. Under its chairman Peter Raven, the committee sought to present the state of the art in biology, focusing on the technical and conceptual developments that have so greatly increased the pace of biological research. At the same time, the committee pointed out—even if it could not always discuss in detail—the many links between scientific issues and ethical issues in biology. Advances in biological understanding have a direct impact on the diagnosis and treatment of disease, theories of human thought and behavior, and ideas about humanity's history and future. So in selecting a topic from the life sciences for the Beckman Center symposium, the intersection between biological research and human values was an obvious choice.

The Board on Biology, under its chairman Francisco Ayala and director John Burris, set up an ad hoc committee to begin planning the symposium. Consisting of Francisco Ayala, John Dowling, Harold Morowitz, Michael Ruse, and Malcolm Steinberg, the committee sketched out the program for the symposium and began contacting potential speakers. Meanwhile, the Board approached the National Research Council for funds to convert the basic materials presented at the symposium into a book. In this, the Board was extending an effort that the Academy complex has made in recent years to disseminate its work to a broad audience by publishing a series of books written for nonscientists.

The symposium brought together many of the leading biological researchers and ethicists in North America, both as presenters and attendees. As the list of speakers shows, four broad areas were covered during the two days of presentations: genetics, development, neurobiology and behavior, and evolution and diversity. In general, the first two speakers in each session addressed the more technical aspects of biological knowledge, while the final speaker addressed ethical issues relating to that knowledge.

Genetics and Humankind

Paul Berg, Stanford University
Leroy Hood, California Institute of Technology
Maxine Singer, Carnegie Institution of Washington

Development and Health

Marc Kirschner, University of California, San Francisco
Barry Pierce, University of Colorado
Arthur Caplan, University of Minnesota

Neurobiology and Behavior

Vernon Mountcastle, Johns Hopkins University
Fernando Nottebohm, Rockefeller University
Michael Ruse, University of Guelph

Evolution and Diversity

William Schopf, University of California, Los Angeles
Francisco Ayala, University of California, Irvine
Edward Wilson, Harvard University

Each of the final three sessions concluded with a panel discussion—moderated by Malcolm Steinberg, John Dowling, and Harold Morowitz, respectively—in which speakers took questions and elaborated on their presentations. Alan Walker of Johns Hopkins University contributed a lively after-dinner speech on human evolution. Peter Raven concluded the symposium by summing up the remarks of the speakers and placing them in a global context.

This book was written using a variety of materials, including the transcripts of the symposium, the published books and articles of the speakers, and additional interviews with the speakers. Though it takes its overall form from the symposium, it departs from it in several ways. In particular, it includes enough background information to be read by relative newcomers to biology. As such, it seeks to present some of the leading ideas of biological research and ethical thought in a clear, nontechnical fashion.

All of the speakers at the symposium gave generously of their time, both before the symposium and after. Within the Commission on Life Sciences, Frances Walton and Kathy Marshall provided the administrative help essential to organizing and holding the symposium. Several people at the National Academy Press worked hard to see the book into print. Michael Edington edited the manuscript and shepherded it through the production process. Dawn Eichenlaub and Francesca Moghari designed the book and its cover.

Two people deserve special mention. John Burris, director of the Board on Biology and executive director of the Commission on Life Sciences, was the guiding light behind the project from beginning to end. Without his boundless energy and good judgment, this book would never have been possible. Finally, Betsy Turvene, executive editor of the National Academy Press, offered the author invaluable encouragement and advice throughout the writing and publication of this book, as she has on many previous occasions.

Steve Olson
Washington, D.C.

*T*o Arnold and Mabel Beckman for their vision in establishing the Beckman Center and for their interest in the ethical issues affecting science, technology, and medicine.

Contents

Introduction

Biology is the most intimate of the sciences. It deals with some of life's most wondrous occurrences, with reproduction and birth, with death and disease, with human abilities and limitations. It seeks scientific explanations for these things, an understanding of how they occur and often why they occur. But explanations do not eliminate the wonder. No matter how much we know about being human, we will always be human.

This book describes ground-breaking research in four of biology's most vibrant fields: genetics, development, neuroscience, and evolution. All of these fields have very long histories, extending back to early human myths and philosophies. At the same time, all of them have experienced tremendous advances in the past few decades. Major questions remain unanswered, but the progress that has been made is in many cases astonishing.

The rapid accumulation of new knowledge in biology has fundamentally changed our views of ourselves and the world around us. By revealing the interconnections between human beings and other organisms, it has reshaped not only science but philosophy and religion. Biological research has also given us an increasing ability to manipulate the biological world, and with that ability has come the necessity of making choices. It has often been a short step in biology from the purely scientific to the ethical, from what one knows or can do to what one believes or ought to do. Progress in genetics, development, neuroscience, and evolution has generated difficult and sometimes agonizing ethical dilemmas, as discussed in each of this book's chapters. Biological research has also raised many broader ethical issues, which form the subject matter for the essays that follow each chapter.

The Molecular Revolution

The most important development in biology in the past half century (though far from the only important development) has been the discovery of the structure and basic function of DNA, the molecule responsible for the continuity of life. This profound scientific achievement sparked the explosive growth of what has come to be known as molecular biology—the study of biological processes on the molecular level rather than the level of cells, organs, organisms, or groups of organisms. Using molecular biology, researchers can trace complex biological processes back to their molecular roots. They can examine the fundamental constituents of biological systems and explain those systems in fundamental terms.

In the 1970s, molecular biology experienced its most significant advance since the uncovering of DNA's structure and function in the 1950s. Researchers learned how to manipulate DNA on the molecular level. Before the 1970s, molecular biologists were largely limited to working with the DNA provided to them by nature. Researchers could cross two interbreeding organisms to get new combinations of DNA, and they could perform a few relatively clumsy operations on DNA using viruses and bacteria. But their inability to control the structure of DNA at will was a major obstacle to continued rapid progress.

That all changed in the 1970s. Using the new techniques of recombinant DNA, biologists could isolate virtually any segment of existing DNA and combine it with other segments of DNA. They could then insert these novel genetic combinations into bacteria, yeast, or even plant or animal cells. They could copy, or clone, DNA segments many times over, creating virtually unlimited amounts of those segments for use in research or other purposes. At the same time, they were developing techniques to sequence pieces of DNA or custom build DNA fragments from simple chemical building blocks. Molecular biology had become as much a technology as a science.

All four of the fields discussed in this book have been influenced to a greater or a lesser degree by the growth of molecular biology. Genetics (Chapter 1) has been transformed by the knowledge that the elements of heredity consist of messages encoded in DNA. The development of a single fertilized egg into a complex organism (Chapter 2) has come to be seen as the playing out of genetic messages. Genetic controls also shape the basic structure and interconnections within the brain (Chapter 3). And over geologic time the messages in DNA have evolved and become more elaborate to produce the overwhelming diversity of life on earth (Chapter 4).

However, not everything in biology can be explained at the molecular level. Many biological processes occur at organizational levels above that of molecules. It is not now possible to describe the development of a human embryo, the higher, more abstract functions of the human mind, or the interactions of organisms in an ecosystem entirely in molecular terms. Current explanations call instead upon the interrelations among cells, or among organisms, or among groups of cells or organisms. Perhaps it will someday be possible to explain these phenomena at the level of molecules (though there are good reasons to think that it may never be possible). But for now, if progress is to be made, explanations must be sought largely at other levels.

The emphasis on molecular biology has also tended to overshadow other developments in biology that have led to important advances. New methods of visualizing biological systems, such as imaging techniques and video technology, have revealed key information about the structure and dynamics of living things. The study of animal behavior has resulted in a much better understanding of how organisms interact with other living things and with their environments. And data storage and computer science have helped researchers make use of the huge amounts of information generated by biological research.

Common Themes

Although the subjects discussed in this book range widely across biology, several ideas emerge over and over. Perhaps the most important concerns the unity of life, as embodied most succinctly in the universality of the genetic code. All organisms, from plants and animals to the simplest bacteria, use the same molecular mechanisms to read the messages encoded in DNA. This is what makes it possible to extract a piece of DNA from a human cell, insert it into bacteria, and have the bacteria read that DNA just as a human cell would.

Plato quotes Protagorus as saying that "man is the measure of all things," but the unity of biology belies that notion. Among the experiments described in this book are ones on fruit flies, toads, mice, monkeys, and canaries. All of these organisms provide insights into human biology, just as the study of human biology reflects on the biology of other organisms.

Another common theme is the increasing unification and interrelation of biological fields. In the 1950s, for example, brain science consisted of three quite separate fields: neuroanatomy (the study of the brain's parts), neurophysiology (the study of the brain's function), and neurochemistry (the study of the brain's chemistry). Then, in the 1950s and 1960s, electron microscopy revealed that there are gaps known as synapses between most nerve cells, and electrical recording of individual cells showed that nerve signals are related to the chemical events that go on in these gaps. The integrated field of neuroscience has been a direct result of this linkage between the brain's structure, function, and chemistry. Similar mergers and cross-fertilizations are going on throughout biology.

A final common theme is the accelerating pace of biological research. Knowledge builds on knowledge, so that all scientific disciplines undergo periods of virtually exponential growth. If any science is now undergoing such growth, it is biology. But growth is not an unmitigated blessing. It can require that difficult ethical choices be made quickly and without a great deal of prior experience. In a fast-moving area of science, the development of ethics can be hard-pressed to keep pace.

This book does not cover many of the other exciting areas in biology, such as immunology or ecology. But by choosing fields that span the range of biology and by focusing on specific research within those fields, this book seeks to give both a broad overview of modern biology and a very real taste of how biological research is done.

1 Genetics and the Human Genome

T he questions are as old as humanity. Why do children resemble their parents? What is responsible for a person's blond hair, green eyes, stocky build? Why do certain diseases, including psychological diseases, run in families?

Before the advent of molecular biology, geneticists approached such questions largely through the study of whole organisms. They bred plants and animals with different traits and observed how those traits appeared in offspring. In sexually reproducing organisms, geneticists knew that these traits had to be inherited from something in egg and sperm cells, and toward the end of the nineteenth century their suspicions began to focus on the chromosomes—long spindly objects seen through the microscope in the nuclei of dividing cells. But for many years the exact nature of chromosomes remained a mystery.

In the 1940s, 1950s, and 1960s, it became clear that the genetic information in each chromosome is carried in a long strand of deoxyribonucleic acid, or DNA. The order of four simple molecules known as genetic bases along the strand specifies an organism's genes, its basic units of inheritance (see box, pages 6–7). But only with the development of recombinant DNA in the 1970s could the nature of complex chromosomes be completely unraveled. The new genetic techniques allow researchers to read an organism's genome, the full complement of its DNA, with unprecedented facility. For the first time, biologists have potentially unlimited access to the information that dictates the structure and function of all living things.

Recombinant DNA and other new molecular techniques have "profoundly altered the practice of biology and medicine," according to Leroy Hood, professor of biology at the California Institute of Technology. Not only have they changed the kinds of studies that are being done, but they have greatly increased the rate at which studies are being done. New findings in genetics and molecular biology are emerging at an unprecedented clip, and at least for the foreseeable future the rate of advance is only going to increase. "I strongly believe that we will learn more about fundamental human biology in the next 20 years than we have in the last 2,000," says Hood.

At the same time, new biological knowledge is raising "a host of perplexing ethical, social, and legal questions," Hood points out. Soon it will be possible to diagnose a person's susceptibility to many common diseases, including cancer and heart disease, generating thorny issues about how that information should be used. Specific genetic defects will become increasingly detectable, forcing people to make unprece-

··

From Genes to Proteins

DNA is the jewel in the crown of molecular genetics. In structure, the DNA molecule resembles a set of railroad tracks twisted lengthwise into a spiral or helix. The rails consist of sugars and phosphate molecules. The ties connecting the rails consist of weakly linked pairs of four chemicals known as genetic bases: adenine, guanine, thymine, and cytosine. Adenine (A) always pairs with thymine (T), while guanine (G) always pairs with cytosine (C).

The double helix structure of DNA is elegantly simple. Its complexity lies in its length. Within each human cell (except red blood cells, which expel their DNA, and sex cells, which contain only half the normal complement of chromosomes) are about 6 billion base pairs of DNA. If DNA actually were the size of railroad tracks, each cell would contain enough DNA to go around the world over 100 times.

Genes are linear sequences of base pairs within DNA. When a gene is being expressed, the DNA zips apart and a ribo-

nucleic acid (RNA) copy is made of its genetic message (see figure). RNA is very similar to DNA, but it uses a different kind of sugar as a backbone and substitutes the chemical uracil (U) for thymine. Through a variety of metabolic steps, this RNA directs the construction of proteins, which consist of chains made up of 20 different amino acids. There is a strict correspondence between the sequence of bases on RNA and the order of amino acids in a protein. Therefore, by knowing the sequence of bases in a gene, it is possible to derive the order of amino acids in a protein, and vice versa (though there are some ambiguities when one is working in reverse).

If DNA is the blueprint for living things, proteins are the bricks and mortar from which living things are built. They give skin and bone its texture, they cause the contraction of muscles, they carry small molecules through the body, they combat foreign substances that enter the body, they generate and transmit nerve impulses, and they control the growth and differentiation dented decisions about how they should live their lives. Techniques are being developed to alter the genetic endowment of a subset of a person's cells to correct inherited diseases. These new capabilities will pose difficult questions for society, questions that reach to the core of what it means to be human.

Genetics and Disease

"Let me begin with an assertion," says Paul Berg, professor of biochemistry at Stanford University and director of Stanford's Center for Molecular and Genetic Medicine, "that all human disease is genetic in origin, or, more accurately, that most diseases are the result of interactions between our genes and our environment." The cases for which this is most obvious are the diseases involving gross disturbances in the number or arrangement of a person's chromosomes. The classic

. . . .

of cells. Most important, in the form of enzymes, they control nearly all of the chemical reactions that occur in living things, including the chemical reactions that convert the food we eat into more DNA and more proteins.

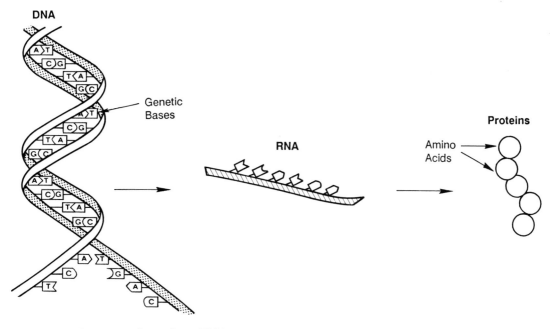

Genetic information flows from DNA through RNA intermediaries to proteins.

example is Down syndrome, in which an extra copy of one of the smallest human chromosomes causes mental retardation, congenital heart defects, and increased susceptibility to infection. Other diseases of this type arise from additional copies of other chromosomes, missing chromosomes, or chromosomes in which the genetic message has become garbled through insertions, deletions, or other obvious rearrangements.

Another category of hereditary diseases is made up of those diseases associated with a defect in a single gene. These diseases are generally divided into two groups, dominant and recessive, depending on their mode of inheritance. The DNA in human cells is divided into 46 chromosomes organized into 23 pairs (Figure 1-1). Each pair consists of a copy of one of the 23 chromosomes in the father's sperm cell and a copy of one of the 23 chromosomes in the mother's egg cell. With the exception of the X and Y chromosomes in males, the two members of each pair are very similar. They contain virtually the same genes in virtually the same order. But the pairs of genes are not identical. Over 30 percent of the corresponding genes in a chromosome pair differ in some way, reflecting the overall genetic variability of the human population.

A dominant genetic disease occurs when an individual receives a defective gene from either parent. In other words, the presence of a functioning gene on one member of a chromosome pair is not enough to overcome the effects of the defective gene on the other member. An example is familial hypercholesterolemia, in which an inability to clear cholesterol from the blood can cause children to suffer fatal heart attacks as early as 18 months. A person with a dominant genetic disease has a 50-50 chance of passing it on to a child.

Recessive genetic defects require that a person inherit defective copies of a gene from both parents. A person who has one defective and one functioning version of such a gene, known as a carrier, suffers little or no ill effects. But each child of two carriers has a one in four chance of inheriting two defective genes and suffering from the disease. Examples of recessive diseases are sickle cell anemia, cystic fibrosis, Duchenne muscular dystrophy, and Tay-Sachs disease.

Over 3,000 single-gene, or monogenic, diseases have been identified. Although individually rare, together they account for a great deal of human suffering. They affect more than 1 percent of liveborn infants and cause almost 10 percent of deaths among children.

The role of genetics is not as clear-cut in multigenic diseases, those involving the interactions of a number of genes with the environment. Examples of multigenic diseases include "hypertension, schizophrenia, manic depression, juvenile diabetes, heart diseases, rheumatoid arthritis, and a host of others," according to Berg. Biologists know that these diseases can be inherited, because they can cluster in families. But they do not yet know the number or identity of most of the genes involved in these diseases.

Cancer should also be classed as a genetic disease, Berg argues, even though it is not a hereditary disease. Cancer is caused by defects in the genetic signals that regulate cell growth and reproduction, and human families clearly have genetic predispositions to certain kinds of

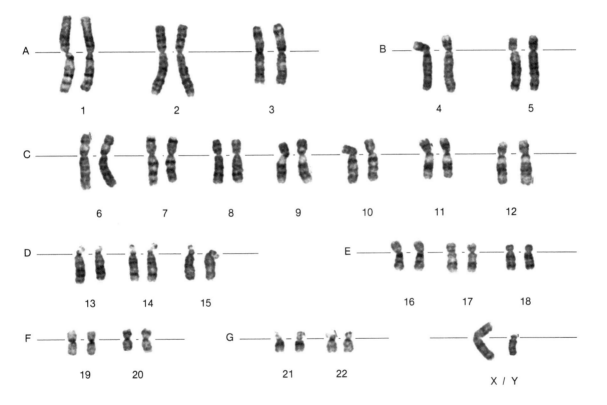

FIGURE 1-1 The DNA in human cells consists of 23 pairs of chromosomes. Twenty two of these pairs, which are numbered roughly in order of descending length, have very similar members, reflecting the equal genetic contributions of mother and father. In addition, each cell carries either two X chromosomes (in the case of a female) or an X and a Y chromosome (in the case of a male). Chromosome photographs such as the one shown here, known as karyotypes, are produced by arresting a cell during division (the chromosomes have doubled but not split apart), staining the chromosomes, and arranging photographs of the chromosomes by size. Karyotype courtesy of Dr. Patricia N. Howard-Peebles, Genetics & IVF Institute, Fairfax, Virginia.

tumors. Also, genetic defects in the human immune system, in DNA repair systems, and in the ability to metabolize carcinogens are associated with a higher frequency of certain tumors.

Even infectious diseases, according to Berg, can be seen as genetic diseases. But in this case, the genes of the infecting organism and their relationship with the host and with the environment determine the course of the disease.

Tracking Disease to Its Source

Molecular biology has already demonstrated the amazing precision it can offer in analyzing genetic diseases. Research has shown, for instance, that sickle cell disease, which is characterized by anemia, impaired growth, and increased susceptibility to infection, is caused by a change in a single genetic base. This change leads to the substitution of one amino acid for another in hemoglobin, the protein that carries oxygen in the blood, causing the molecule to misfunction when oxygen is scarce. Another example is phenylketonuria (PKU), which can cause severe mental retardation if undiagnosed; it is caused by a defect in the gene coding for an enzyme that converts one amino acid into another. "Every few months another gene is isolated and the structural defect is identified," says Berg. "It's not too optimistic, I think, to predict that the genes and their defects for many of the monogenic diseases will be known within the next five years."

The impressive accomplishments of the past emphasize how much remains to be learned, however. Of the over 3,000 monogenic diseases now recognized, the responsible gene has been identified in only about 100 cases. The genes involved in multigenic diseases and disease susceptibilities remain largely unknown.

The ideal situation would be to know the gene or genes involved in every human disease, their location on the chromosomes, the nature of the defect associated with the disease, and the way in which the defect contributes to the disease. This information is known for very few diseases, and attaining it for the majority of diseases will take many decades. But biologists are starting to systematically pursue some of the early stages of such a program. They are constructing maps of the human chromosomes that give the locations of known genes. By mapping these locations, researchers can develop genetic probes to determine whether a person has a normal or a defective gene, leading to great advances in the diagnosis and treatment of disease. Genetic maps can also reveal patterns in the way genes are organized and regulated, leading to a greater understanding of how genes function in health and disease.

The scientific community has also been considering a much more ambitious proposal: a plan to determine the exact sequence of the billions of genetic bases making up the human genome. This information would be an invaluable resource for biologists. It would allow them to identify all of the genes within a given region of a chromosome, including many currently unknown genes. Genetic differences among individuals could be compared, revealing a great deal about the functioning of normal and defective genes. The mechanisms that control the expression of genes and the processes involved in development could be deciphered. The structure and organization of genes in different species could be contrasted, leading to insights into the evolutionary processes that resulted in those species.

The complete sequence would not answer every question in biology. It would not establish exactly how genes are controlled or how gene products function within a cell or organism. It would not completely explain how people are different or how humans have evolved. But it is unlikely that these questions can be answered without a deep understanding of the human genome.

Putting Genes on a Map

Geneticists were mapping genes to chromosomes well before they knew how chromosomes are constructed. The first genetic map was made in 1913, for five traits carried on the X chromosome of the fruit fly *Drosophila melanogaster*. Today, maps of human chromosomes are being made using the same principles.

The method used to make such maps involves analyzing how traits are passed down from generation to generation. In a parent's reproductive system, each chromosome pair separates during the formation of egg and sperm cells, reducing the original 46 chromosomes to 23. If this were all that occurred during reproduction, inheritance would be a straightforward matter; chromosomes would remain intact and be passed unchanged between generations. But chromosomes do not remain intact. Rather, before the formation of egg and sperm cells, each chromosome pair can exchange parts through a process known as crossing-over.

Imagine three different genes located on a pair of chromosomes (Figure 1-2). Each chromosome can have different versions of these genes, which interact according to the usual rules for dominant and

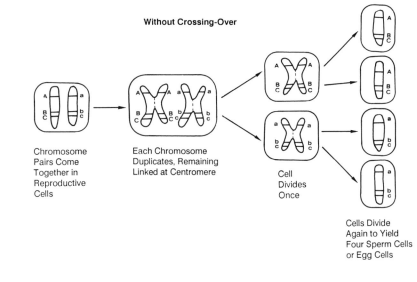

Without Crossing-Over

Chromosome
Pairs Come
Together in
Reproductive
Cells

Each Chromosome
Duplicates, Remaining
Linked at Centromere

Cell
Divides
Once

Cells Divide
Again to Yield
Four Sperm Cells
or Egg Cells

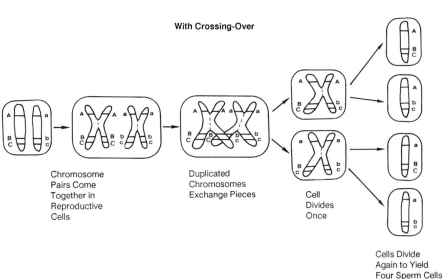

With Crossing-Over

Chromosome
Pairs Come
Together in
Reproductive
Cells

Duplicated
Chromosomes
Exchange Pieces

Cell
Divides
Once

Cells Divide
Again to Yield
Four Sperm Cells
or Egg Cells

FIGURE 1-2 During the formation of egg and sperm cells, a process known as crossing-over can unlink versions of genes located on the same chromosome. In the above diagram, *A* and *a* are versions of a specific gene located on a chromosome pair, as are *B* and *b* and *C* and *c*. During crossing-over, *A* becomes unlinked from *B* and *C* in two of the four sperm or egg cells produced. If the resulting cells produce offspring, the genetic linkage will differ between parent and child. This information can be used to deduce the relative distance between the different genes.

recessive genes. For instance, gene *B* in Figure 1-2 may be for brown eyes—a dominant trait—while gene *b* is for blue eyes.

If crossing-over and other forms of genetic recombination never occurred, the versions of genes located on a given chromosome would always be inherited together. In genetic terms, they would be permanently linked (top half of Figure 1-2). But crossing-over can reshuffle the genes on a chromosome pair, resulting in new genetic combinations.

The trick in genetic mapping is to observe how often this process separates the versions of genes on a chromosome. As shown in the bottom half of Figure 1-2, genes that are close together tend to stay together, whereas distant genes are easier to separate. By noting the frequency of separation, researchers can calculate the distance between genes and assign them relative locations on a chromosome.

Another way to map genes draws on the appearance of chromosomes under a microscope. If the chromosomes are stained during cell division with certain chemicals, alternating bands of light and dark regions appear, with up to several dozen bands on a single chromosome (Figure 1-3). These cytogenetic bands distinguish the chromosomes and provide broad landmarks along their length. But they are still quite large, with each band containing an average of 100 genes.

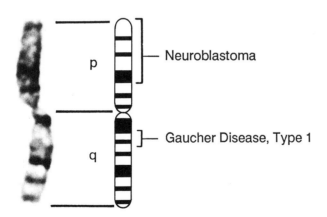

FIGURE 1-3   Several dozen disease-causing genes have been mapped to human chromosome 1. The gene responsible for Gaucher disease, type 1, a recessive disease characterized by an enlarged spleen, skin pigmentation, and bone lesions, has been mapped to a section of the q arm of the chromosome. One or more genes that control the spread of tumors in neuroblastoma, a cancer of nerve cells, have been localized to part of the chromosome's p arm. Karyotype of chromosome 1 courtesy of Dr. Patricia N. Howard-Peebles, Genetics & IVF Institute, Fairfax, Virginia.

Some people with genetic disorders have recognizable abnormalities in their cytogenetic banding patterns. For instance, the gene for Duchenne muscular dystrophy was mapped to a specific region of the X chromosome by noting that some sufferers of the disease were missing that portion of the chromosome. Translocations, in which part of one chromosome has broken away and become attached to another chromosome, and fragile sites that are susceptible to breakage have also pointed to the locations of specific genes.

If the base sequence of the gene being mapped is known, it can be located on a chromosome by making radioactive copies of part or all of the gene (by synthesizing or cloning the gene using radioactive constituents). These DNA probes can then be mixed with human chromosomes under chemical conditions that cause the DNA strands to temporarily separate. When a single-stranded probe encounters a matching single strand of DNA on a chromosome, the two combine or hybridize to produce double-stranded DNA. The radioactivity given off by the probe then functions as a marker to find the chromosome and the approximate location of the gene.

Another method of locating genes on chromosomes involves an intriguing technique known as somatic cell hybridization. When human cells and mouse tumor cells are grown together under the proper conditions, they tend to fuse and form hybrid cells. As these hybrid cells grow and divide, they lose most of their human chromosomes. But often one or a few human chromosomes will become stably established in a particular cell. The result is a mouse tumor cell line containing specific human chromosomes. By looking for human proteins produced by a given gene, that gene can be assigned to a chromosome carried in a cell line. Genes with known sequences can also be located using DNA hybridization.

All of these techniques can establish the relative or approximate location of a gene on a chromosome. But to map a gene to an exact chromosomal location requires the techniques of recombinant DNA.

A Genetic Scalpel

The ability of biologists to recombine DNA at will originated in the discovery of enzymes in bacteria that can cut DNA at specific sequences of four to ten genetic bases. These so-called restriction enzymes, which bacteria evolved to fight invasions of foreign DNA, were a marvelous gift to biologists. They allow researchers to slice DNA at specific locations, so that large DNA molecules can be cut into smaller, more manageable pieces. Then, using a class of enzymes known as ligases, biologists could splice together any two pieces of DNA formed with the same restriction enzymes. In this way, they could create mosaics of DNA, known as chimeras, that have never before existed in nature. (In Greek mythology, the chimera is an animal with the head of a lion, the body of a goat, and a serpent for a tail.)

Recombinant DNA has transformed the mapping of the human genome. It has enabled researchers to cut complex genomes into pieces, each of which can then be cloned many times over. These pieces can be organized into a library, creating a complete collection of all the DNA in an organism. Individual volumes in the library can be further analyzed to map specific genes or pieces of DNA. The logical conclusion of this process is the next step beyond mapping: the sequencing of genetic bases in part or all of an organism's genome.

One of the products of recombinant DNA technology has been a family of genetic markers much more precise and powerful than the banding patterns on chromosomes. Human beings are much more alike genetically than they are unalike. Only about one in every hundred base pairs of DNA are different between any two people. Many of these differences have no effect on the functioning of the genes, but others contribute to the crucial differences that make us unique: differences in physical appearance, aspects of personality, susceptibility to disease. Without these genetic differences, people would all be as alike as identical twins.

Using recombinant DNA, genetic differences among individuals can also serve as markers along human chromosomes. First DNA from a person's cells is cut into pieces using a specific restriction enzyme, and the resulting pieces are placed along the edge of a gel. Under normal conditions, DNA has a slightly negative electric charge, so when an electrical potential is applied across the gel the DNA is attracted to the positive charge on the other side of the gel. The smaller pieces of DNA move faster through the gel than do the larger ones. When the DNA pieces are spread across the gel, this process, known as electrophoresis, is stopped. The result is a virtually continuous series of bands, with each band corresponding to a DNA fragment of a different length. Researchers can then highlight specific bands using radioactive DNA probes that combine with given DNA sequences.

If two individuals have a difference, known as a polymorphism, in the region of DNA being tested, that difference can cause the restriction enzyme to produce DNA fragments of different lengths. For instance, say a person has a difference in a sequence recognized by a restriction enzyme, causing the sequence to remain uncut (Figure 1-4). If so, the

DNA fragments from that person will have different lengths than fragments from a person without that polymorphism. The difference in the DNA responsible for the different fragment sizes is called a restriction fragment length polymorphism or RFLP (pronounced "riflip").

The presence or absence of a particular RFLP in a person's DNA can act as a genetic marker on a chromosome. As with entire genes, it can be tracked from generation to generation and used in linkage studies. If a RFLP is located close to a defective gene, it will tend to be inherited with that gene, with the degree of linkage depending on the actual distance between the RFLP and the gene. If the identity of the gene is unknown—as is still the case for the great majority of genetic diseases—the RFLP can serve as a surrogate marker for the gene.

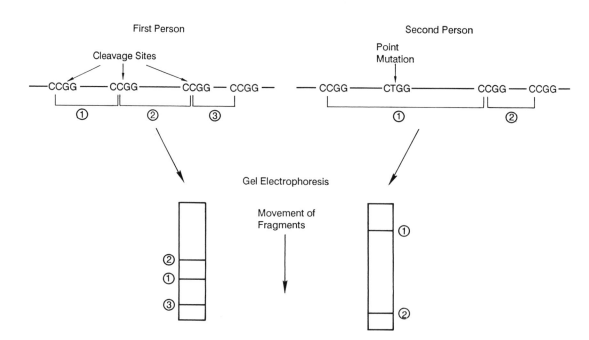

FIGURE 1-4 A point mutation at a single genetic base can keep a restriction enzyme from cleaving a genetic location that is cleaved in another individual. This results in DNA fragments of different lengths, which can be separated and distinguished using electrophoresis. Such differences, or polymorphisms, which can also arise from genetic additions, deletions, or translocations, can serve as markers of specific chromosomal locations.

RFLPs and Disease Diagnosis

The implications for genetics of RFLPs are "extraordinary," according to Berg. Geneticists have already located hundreds of RFLPs scattered throughout the human genome. At first, a RFLP may be used as a flag to indicate the presence of a disease-causing gene closely linked to the RFLP. Later, RFLPs may be used to track down the actual location of that gene, so that the nature of its defect can be determined.

The use of RFLPs and other genetic probes to diagnose genetic diseases "will change medicine in a very profound way," according to Hood. For instance, it will eventually be possible to detect the genes responsible for virtually every monogenic disorder. These genes could be detected after birth or prenatally, giving parents an option to terminate a pregnancy. Carriers of defective genes could also be identified, allowing them to decide whether to have children and risk passing on the disease.

The identification of carriers and prenatal testing could greatly reduce the toll that many monogenic diseases take on the human population, Berg points out. An example is Tay-Sachs disease, a recessive disease occurring most frequently among Jews of European descent that causes retardation, paralysis, and early death. A test for Tay-Sachs disease that can identify carriers and affected fetuses has been available for a number of years. This test "has virtually eliminated Tay-Sachs disease from the Jewish population," Berg says.

As scientists learn more about the role of genes in multigenic diseases, it will increasingly be possible to determine an individual's susceptibility to such diseases. In time, diagnostic tests should be available to determine a person's susceptibility to such common diseases as cancer, heart disease, and diabetes.

Such tests could provide "a new opportunity for preventive medicine," says Hood. Now, genetic testing is confined largely to fetuses and expecting parents. But in the future it will be possible to diagnose the disease susceptibilities of anyone and enter that information into a medical record. That information could help a person to avoid certain habits, diets, or environmental conditions that might lead to the disease. New therapeutics might also be developed that could lower the chance of contracting a disease when a susceptibility exists. "We can start to think about changing some fundamental aspects of human health care," says Hood.

As an interesting sidelight, RFLPs and other DNA probes have also assumed a growing role in forensic medicine. The patterns in a person's DNA are just as distinctive as a fingerprint, and in some crimes DNA samples are easier to find than fingerprints. Already, genetic tests have been used to identify rape suspects, murder suspects, and the parents of children applying for immigration. "Very soon we will be in a position to translate genetic patterns into electronic databases and compare particular DNA fingerprints with any others that come up," says Hood. "This raises challenging social and ethical questions."

The Price of Knowledge

The social and ethical questions posed by genetic analysis extend far beyond DNA fingerprinting. As testing for diseases and disease susceptibilities increases, it will be possible for people to learn a great deal about themselves and their futures. If it were possible for a person to know what he or she was most likely to die of, would that person want to know? Today, genetic testing is often done as the prelude to some therapeutic intervention, from the abortion of a severely affected fetus to dietary or pharmacological intervention. But treatment may not be possible for some of the diseases that will be detectable in the future, at least until researchers apply new understandings of genetic information to the development of new therapies.

A stark example of the dilemma posed by the gap between diagnosis and treatment is Huntington's disease, which threatens about 125,000 people in the United States. The disease is caused by a dominant gene inherited from either the mother or the father; therefore, a person with an affected parent has a 50-50 chance of also having the disease. The symptoms of the disease, beginning with loss of control over movement and proceeding to dementia and eventual death, generally do not appear until after the childbearing years. As with many monogenic diseases, there is no treatment for Huntington's disease.

Through extensive analyses of RFLPs from families afflicted by Huntington's disease, geneticists have narrowed the location of the Huntington's gene to a million-base-pair region at the end of chromosome 4. By studying the RFLPs of individual families, researchers can usually find markers that signal the presence of the defective gene with a high degree of accuracy. The question then becomes, Will a person at risk of Huntington's disease want to know whether he or she carries the gene for the disease? If a person is trying to decide whether to have a child and risk passing the gene on to the next generation, the information is clearly useful. But if a person is past the childbearing years or does not want to have children, why undergo a genetic test that has the potential to predict future illness and early death?

As genetic testing becomes more powerful, similar dilemmas will emerge. The ability to test for the presence of a gene is an inevitable step in the process of understanding a disease and developing ways to treat it. But some diseases will always remain untreatable, and for others the ability to diagnose a disease is imminent whereas the treatment for it is much less certain.

Information from genetic testing will also raise a number of contentious social questions. Should employers or insurance companies have access to such information, even though in some cases it could lead to loss of a job or loss of insurance coverage? What kinds of traits or conditions should prospective parents be able to test for in deciding whether to have children or abort a pregnancy? Guidelines will have to be established for the use of such information to preserve individual confidentiality and autonomy.

Increased genetic testing will also require that people understand the information they are being given. An increased susceptibility to a disease does not mean that a person will inevitably get that disease. In fact, the chance of getting the disease may be quite small. Yet such a diagnosis could greatly increase a person's anxiety and affect important decisions.

Even in cases in which a diagnosis is more certain, the potential for misinterpretation will be large. People who have a disease caused by a single gene can show enormous variations in the manifestations and severity of the disease. This variation is so great, says Berg, "that some physicians who see these patients cannot accept that they're all caused by the same genetic defect." In part, says Berg, these differences are caused by a person's broad genetic background, which probably will not enter into the diagnosis. They are also caused by the environmental influences a person experiences, which can be endlessly variable. People will therefore have to understand the full range of outcomes that could accompany a specific diagnosis.

These are not new issues in human genetics. Similar situations have already been addressed by researchers, clinicians, patients, and policymakers. But the issues will become more numerous and widespread as biologists learn more about the connection between a person's genetic heritage and disease.

From Mapping to Sequencing

From a purely scientific standpoint, biologists generally agree that a program to map the human genome through the construction of a large library of RFLPs and other genetic markers is a worthwhile goal. But the new techniques of molecular biology allow more to be done. Within a few years it should be possible to begin systematically sequencing the 3 billion base pairs in the human genome. It would be the largest project ever undertaken in biology, with characteristics much different than those of normal biological research. There is much less consensus among biologists about how to undertake such a program.

One thing is clear, however: it will be a formidable task. Apart from the technical difficulties involved in sequencing, the project will produce a tremendous amount of information to store, analyze, and disseminate. There are about 3 billion bases that would need to be sequenced in the human genome (technically, only one member of each chromosome pair needs to be sequenced, since the members of a pair are so similar) If every letter in this book represented one of these bases, it would take roughly 10,000 of these books to represent the entire sequence. Furthermore, sequencing programs will have to acquire sequences from different individuals and from other species to make the best use of the human sequence. So far, only about 2 million base pairs of the human genome have been sequenced and stored in a central data

· ·

Building a Better Sequencer

The standard laboratory method used to sequence DNA involves labeling fragments of the unknown DNA with a radioactive marker and passing the fragments through four different electrophoretic gels, one for each kind of base. Once the fragments have been separated on the gels, it is possible to read the base sequence from the order in which the fragments appear.

The problem with this method is that it is tedious and expensive, requiring skilled scientists and the use of hazardous chemicals and unstable radiosotopes. If sequencing the entire human genome is to be practical, sequencing methods must become more efficient.

One way to increase efficiency has been developed by Leroy Hood and his colleagues at the California Institute of Technology and Applied Systems, Inc. Instead of labeling each DNA fragment with the same radioactive marker, the fragments are labeled with four distinct flourescent markers, one for each kind of base. This makes it possible to run all of the fragments through a single electrophoretic gel. As fragments of different size move down the gel, a laser causes the passing bands of DNA to fluoresce, and the emitted light is recorded by a computer (see figure). Since each flourescent marker produces a different color, the computer can read the DNA sequence from the passing DNA bands.

This technique offers a number of advantages over the conventional sequencing method. For one thing, it reduces the errors inherent in using four different gels. Also, because the sequencing is automated, the process proceeds much more quickly. Commercial versions of the device can sequence about 6,000 bases per day, or approximately 1 million bases per year.

base, less than 0.1 percent of the total. Without a special effort to sequence the human genome, it will not be done for many years, if ever.

There are several broad objections to a sequencing program that must be answered for the project to proceed, according to Hood. The first asks whether obtaining the full sequence would in fact be scientifically uninteresting and therefore a misallocation of resources. Hood agrees that much of the mapping and sequencing would be routine, repetitive work. Ideally, he says, much of it can be automated (see box, pages 20–21). It is the sequence itself that will generate "the staggering and exciting kinds of scientific endeavors." Also, Hood questions whether the United States can afford *not* to undertake a sequencing program. "I would argue that the impetus provided by this program is going to have a major impact on whether the United States stays competitive in the industrial biotechnology area."

A second objection is that such a project will draw funds from other areas of biology, and Hood finds this to be "a much more serious objection." But the technologies and information developed through a sequencing program will find applications throughout biology, Hood

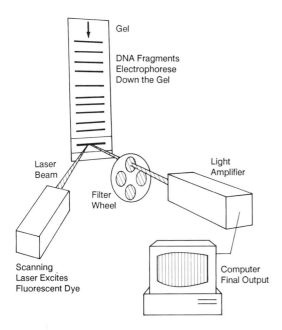

Gel

DNA Fragments
Electrophorese
Down the Gel

Laser
Beam

Light
Amplifier

Filter
Wheel

Scanning
Laser Excites
Fluorescent Dye

Computer
Final Output

By using four different fluorescent markers to label the DNA fragments, the fragments can be run on a single gel and detected by a laser beam linked to a computer.

points out. He also believes that the project will be "sufficiently compelling" to generate new sources of funding for itself. Several committees have studied sequencing proposals and have concluded that funding in the neighborhood of $200 million per year would be appropriate for such a program. While this is a substantial amount of money, it is only about 3 percent of the total amount of funds spent by the federal government on biological research each year.

A third objection is that the project will be "big science" and therefore at odds with the traditional approach to biological research, in which most work is done by small, independent groups of scientists. But Hood contends that "it is not big science in the same sense that projects such as the superconducting supercollider and the space shuttle are big science." The costs of the instruments required are modest, and the instruments can be widely disseminated. Research proposals for aspects of mapping and sequencing will be peer reviewed, and this research will also be widely distributed. Furthermore, once the information is available, it will greatly increase the power and range of what small groups can do, and they will no longer need to spend great amounts of time on routine mapping and sequencing.

The final objection is that the technology is not yet adequate to do the job, and this is the one objection that Hood finds convincing. As currently performed, sequencing is tedious, time-consuming, labor-intensive, and expensive, according to Hood. "Technology development is what we really should concentrate on," he says. "This is an area of major deficiency at this time."

Sequencing the human genome will involve building up a library of overlapping pieces of the genome, which can be stored in a central location, easily reproduced, and sent to investigators to be sequenced and studied. This is now relatively easy to do with small pieces of DNA. They can be cut with restriction enzymes, inserted into pieces of bacterial or viral DNA known as vectors, and maintained in culture as clones. The problem is one of scale. Conventional vectors can hold a maximum of about 40,000 bases, meaning that with an average overlap of 10,000 bases, it would take over 100,000 different clones to store the entire human genome. Maintaining a culture library of this size would require an administrative structure much larger than any in existence today. Furthermore, some parts of the human genome are difficult to clone, and there are concerns about the stability of DNA in a culture library.

Progress is being made on many of these problems. Researchers are developing new vectors in yeast that may be able to carry 500,000 to 1 million DNA bases, over ten times the size of the fragments that can be maintained with current systems. Instead of needing over 100,000 clones to encompass the entire human genome, it could be done with several thousand. Restriction enzymes have recently been discovered that cut DNA into very large pieces, and a new method of electrophoresis using pulsed electric fields can separate DNA fragments 200 times larger than the maximum possible with conventional electrophoresis. These are the kinds of technological developments that will be necessary to make sequencing the human genome economically and scientifically feasible. (The box on pages 24–25 discusses several other technologies that will be essential to sequencing efforts.)

Hood advocates that an effort to sequence the human genome proceed in stages. During the first stage, new technologies would be developed to increase the efficiency of DNA sequencing five- to ten-fold. At the same time, detailed mapping of the human genome could be under way, which would provide a framework for the sequencing effort. A phased approach would also allow systems to be developed to collect and disseminate DNA clones and to store and analyze the huge amounts of data that will be generated. Once this technological infrastructure is in place, Hood says, the complete sequencing of the genome could begin.

Prospects for Human Gene Therapy

The knowledge provided by mapping and sequencing the human genome may make it possible to achieve one of the most provocative of the new biotechnologies: human gene therapy. Interest in human gene therapy arises from a depressing fact: for the majority of monogenic diseases, no effective therapies are available. For some single-gene defects, dietary restrictions, the use of drugs or biologic agents, or transplantation of tissues or organs may alleviate part or all of the symptoms. But for many such genetic defects there is no alternative to "debilitating and progressive disease leading to suffering and early death," according to Berg.

Any disease to be treated with human gene therapy must meet a number of "very formidable" criteria, Berg says. The gene responsible for the disease and the molecular nature of its defect must be known. The disease must involve cell types that are accessible and well-characterized. The disease cannot begin to exert its harmful effects until after birth, since a variety of technical and ethical constraints prohibit gene therapy before birth. Also, the defect must be limited to a single gene. "We certainly enter the realm of wishful thinking when the therapy aims to modify more than one gene or to rectify the defects resulting from chromosomal abnormalities," notes Berg.

The gene therapy being considered for humans involves placing normal genes into somatic, or body, cells, not into germ line, or sex, cells. In this way, "the therapy stays with the treated patient and is not transmitted to offspring," Berg says, "so any undesirable effects of the integration are not going to be propagated to future generations."

Given the criteria that candidate diseases must meet, only a handful have received serious consideration. The most attention has focused on a rare disease caused by defects in the enzymes necessary for normal immune system development. Children with the disease, known as severe combined immunodeficiency disease, must live in totally sterile

....

Protein Sequencers and Synthesizers

Recombinant DNA technology is not the only driving force behind the rapid advances now occurring in genetics. Another powerful influence has been the development of microchemical instruments that can analyze and synthesize genes and proteins with remarkable proficiency.

One such instrument is known as a protein sequenator. Essentially, this device is a chemical scissors that can clip off one amino acid at a time from the end of a protein and determine its identity. Since the development of the first sequenator in 1967, the amount of a protein needed for sequencing has steadily declined. Today, biologists are on the verge of being able to sequence the proteins purified by the most sensitive technique now available—two-dimensional electrophoresis. This technique separates complex protein mixtures in one dimension by size and in the other dimension by charge (see figure). Within a few years, it should be possible to sequence virtually every protein that appears in a two-dimensional gel. The result will be a dramatic increase in the number of proteins that can be analyzed.

Once the sequence of amino acids in a protein is known, a synthetic gene can be made for the protein by converting the protein's amino acid sequence into the corresponding base sequence. DNA synthesizers can now construct DNA fragments up to 200 bases long, and these fragments can be joined together to make longer sequences. A synthetic gene or part of a gene can be hybridized with chromosomal DNA to locate the original gene for the protein. A synthetic gene or the cloned original gene can also be introduced into bacteria or other hosts to manufacture large quantities of the protein for research or therapeutic uses.

Biologists are also using microchemical instrumentation to explore one of the most fundamental issues in biology: how the sequence of amino acids in a protein determines its structure and hence its function. Using DNA synthesizers, researchers are making synthetic genes that are slightly different from the original genes, resulting in proteins with different amino acids in particular locations. By observing how these modified proteins fold and operate, researchers hope to uncover the general principles that govern the structure and function of proteins. Researchers can also directly synthesize protein units over 100 amino acids long to examine which amino acids "bubbles," since their immune systems cannot protect them from common viruses, bacteria, and fungi.

Researchers have developed an experimental procedure in mice that could be applied to humans if it proves effective and safe (Figure 1-5). First, bone marrow cells are removed from the animals and mixed with an infectious agent known as a retrovirus. When retroviruses infect a cell, they insert a copy of their genetic information into the genome of the host. By genetically engineering the retroviruses to contain a normal version of the defective gene, scientists can insert the normal gene into the bone marrow cells. The cells are then reimplanted into the mice where, presumably, they will produce the missing gene product.

Unfortunately, says Berg, the results to date "have been rather dis-

....

are important for particular functions. In time, such information may make it possible to construct proteins with new and useful properties, inaugurating an era of protein engineering that is likely to have an even greater impact on science and medicine than genetic engineering has had.

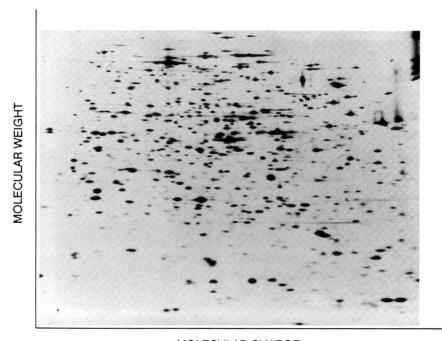

MOLECULAR WEIGHT

MOLECULAR CHARGE

Two-dimensional electrophoresis can separate thousands of proteins extracted from a few hundred cells. Each dot in this separation indicates the presence of a single kind of protein. Photograph courtesy of Leroy Hood.

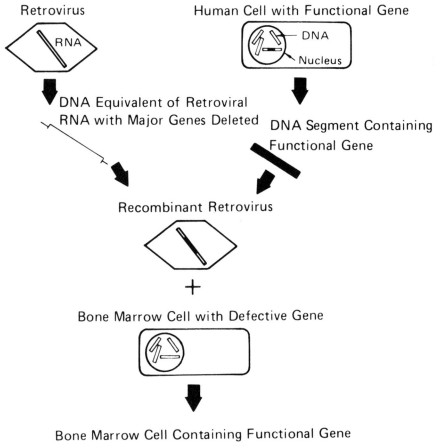

Retrovirus

Human Cell with Functional Gene

RNA

DNA

Nucleus

DNA Equivalent of Retroviral
RNA with Major Genes Deleted

DNA Segment Containing
Functional Gene

Recombinant Retrovirus

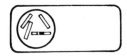

Bone Marrow Cell with Defective Gene

Bone Marrow Cell Containing Functional Gene

FIGURE 1-5 Much of the research on human gene therapy has focused on retroviruses, infectious agents that can insert their own genetic material into the DNA
of cells they infect. The genome of a retrovirus consists of RNA, which is enzymatically
copied into DNA when the virus invades the cell. In the scenario shown above,
involving bone marrow cells with a defective gene, the major genes in a DNA copy
of the retroviral RNA can be deleted and replaced with a functional version of the
defective gene, along with the appropriate regulatory signals to ensure the expression
of the gene. Once the recombinant molecule has been reconverted to RNA, the
bioengineered retroviruses can be used to infect bone marrow cells withdrawn from
an organism with the defective gene. The retroviruses insert the functional gene into
a random location in the cells' DNA, and the transformed cells are reimplanted into
the organism.

couraging and may prompt others to begin looking at other potential vectors." The genes can be inserted into mouse bone marrow cells and made to produce the gene product, but only for a limited time. "For reasons that are totally mystifying at the moment, the gene that has been introduced by the virus is turned off over some varying period of time. Although the animal has acquired the foreign DNA, it no longer expresses the gene of interest." Much effort has gone into trying to maintain the expression of the introduced gene, but without success. "My feeling is that this probably reflects our lack of knowledge about the development of the cell," says Berg.

Human gene therapy has other potential problems that are only beginning to be addressed, notes Berg. One is that the retrovirus inserts its genetic message into the cell's DNA at random and uncontrollable locations. It is possible that the foreign DNA would integrate into the middle of a gene essential for the survival of the cell, which would destroy it. Or the gene could insert itself in such a way that it increases the activity of a gene regulating cellular growth, leading to tumors.

"The inability to control the site at which the vector DNA integrates is one of the major handicaps at the moment in this approach," Berg points out. As a result, he and his coworkers are examining ways to target the introduced DNA to specific locations. Perhaps vectors can be developed that would recognize sequences within the cell's genome and integrate its DNA in a predictable way around that sequence. Even more desirable would be some sort of agent that homes in on the defective gene and somehow repairs it in place. But knowing whether any of these approaches are feasible will require a much more sophisticated understanding of the workings of the human genome.

Science and Scientists from the Public's Perspective

"The end of man is knowledge, but there is one thing he can't know. He can't know whether knowledge will save him or kill him. He will be killed, all right, but he can't know whether he is killed because of the knowledge which he has got or because of the knowledge which he hasn't got and which if he had it, would save him."
—Robert Penn Warren
All the King's Men

When scientists confront the dilemma posed by Robert Penn Warren, they tend to respond optimistically, says Maxine Singer, president of the Carnegie Institution in Washington, D.C. "We believe that the acquisition of new knowledge is both wondrous and good. We also believe that the quest for better comprehension is a fundamental human trait, one which sets *Homo sapiens* apart from other living species. We reinforce this belief by a conviction that original research is a creative endeavor, linked to artistic and literary creativity in method and talent. And we believe that these last considerations say that science is a thoroughly human enterprise, that through science we are expressing human as well as humane values."

But the public's perception of science can be ambivalent and often pessimistic, Singer acknowledges. "For the most part, nonscientists find the continuing quest for knowledge somewhat frightening," she says. To some extent, this fear extends even to scientists. "Often at Washington cocktail parties, where everyone asks strangers, 'What do you do?' I find that the answer 'molecular biologist' is likely to drive the questioner to the far end of the room." The same attitude surfaces in the popular media. In television shows and best-selling books, scientists are often depicted as "frightening, usually unsympathetic, almost inhuman."

Despite these widespread impressions, Singer does not believe that the public's view of science is totally negative. Public attitudes are too numerous, diverse, and at times contradictory to characterize one-sidedly. People are eager to make use of the fruits of scientific research, and many avidly follow reports of scientific developments. Biologists and other scientists have also devoted considerable time in recent years to explaining their work to the public. Partly as a result, Singer notes, over 50 percent of all Americans say that they know what DNA is.

But the tension between the acquisition of new knowledge and the fear of that knowledge remains widespread in society. It is a troubling tension to scientists, Singer believes, because scientists in the United States rely on the public for support. On a purely financial level, the majority of scientific research is paid for with public funds. And more broadly, in a democracy, scientific work on controversial subjects can be slowed or halted by public opposition, even if engendered by unwarranted fears.

The Uses of Knowledge

Scientists are well aware of the ambivalence with which the public views their work. One sign of this awareness is the social contract through which scientists solicit funds for research. While scientists pursue knowledge, the public can gain from that knowledge—new treatments for disease, for instance, or agricultural improvements. Current plans to map and sequence the human genome are a good example. One set of rationales for such a project speak of an increased understanding of disease, development of new therapeutic agents, and heightened international competitiveness. But any such benefits will be built on a new base of knowledge about the structure and functioning of DNA.

"There is nothing wrong with these honest arguments" about the practical benefits of science, Singer says. Scientists want to make the world a better place; for some, that may be their primary motivation. But the fundamental purpose of science is to learn more about the world.

The public's ambivalence toward science also emerges in other ways. For instance, new scientific discoveries and their implications are extensively covered by the media. "Indeed, we often read of new discoveries first in the press and only later in journals," Singer points out. But the press also devotes considerable time to scientific controversies that most scientists consider relatively minor or beside the point. Transgressions of scientific standards, whether substantial or insignificant, become front-page news. The views of a small minority may be presented as a counterpoint to widely held scientific outlooks, giving the minority viewpoints a credence that they do not deserve.

The public's apprehension over new knowledge can be particularly acute in biology. Biology seeks to describe the fundamental nature of human beings, offering a self-knowledge that is not always reassuring. The increasing ability of biologists to manipulate biological systems also is heightening the impact of biology on the modern world.

Science and Myth

Much of the public's unease over scientific advances arises because of the way in which science can conflict with long-standing premises, explanations, and authorities, Singer contends. In many cases, scientific explanations of natural phenomena are becoming available where mythologic explanations have traditionally held sway. These conflicting viewpoints influence public debate in a number of ways.

An obvious example is the continuing debate over the teaching of evolution. Modern genetics fully supports the conclusions of evolutionary biologists that human beings evolved from earlier forms of life. Nevertheless, surveys show that over half of all Americans think that biblical creation myths should be taught in science curricula in American schools, even though mainstream religious leaders do not support this view. "Thus it is not only religious fundamentalists who prohibit us from ending this debate," Singer maintains.

Another example comes from the prospect of human gene therapy. In a recent poll, when gene therapy was presented as a means of curing fatal diseases and preventing inherited birth defects, 84 percent of the respondents were in favor of it. But many of this same group also, and inconsistently, said that it was morally wrong to tamper with the genetic code of humans.

Ideas about the origins of disease also reveal the gap between scientific and mythologic explanations of events. Some people in the United States, for instance, believe that AIDS is a form of divine retribution against homosexuals and drug abusers. In this, they echo the views of John Woolman, a prominent American thinker in the 1700s, who wrote, "I have looked on the Smallpox as a messenger sent from the Almighty, to be an assistant in the cause of virtue." "One cannot help but wonder," Singer responds, "how Woolman would have reacted to the recent worldwide eradication of the Almighty's messenger."

Replacing mythologic explanations with scientific ones "will be a difficult job," according to Singer. Many myths are associated with accepted authorities, such as religion, or with unquestioned assumptions, such as the inviolability of nature. Myth tends to be seen as human, Singer says. Science tends to be seen as inhuman.

A Mechanistic View of Nature

If scientists are to succeed in promoting their view of the world, they cannot downplay those aspects of the scientific enterprise that the public finds troubling. Instead, they must work to see that the scientific viewpoint is not distorted by those who oppose it. In particular, says Singer, biologists need to discuss in a straightforward way the profoundly mechanistic view of the natural world that has emerged from their work.

Scientists welcome a mechanistic view of nature, because it means that the world is knowable, and perhaps explicable. But this viewpoint can easily be caricatured to imply that the world is mechanical, devoid of purpose or meaning, inhuman. In fact, says Singer, a mechanistic view of nature by no means rids the world of its significance and beauty. "It is time for us to declare that our respect for nature, our love of its beauty, our concern for the environment, are not diminished by a molecular view of its workings," she says.

A mechanistic view of nature also does not imply that scientists reject the values that some see as inextricably associated with traditional premises and authorities. Scientific understanding does not undermine fundamental human assumptions "about good and evil, about justice, about freedom, about joy and sadness," Singer says. "We need only look at other geneticists to realize that human interactions are pretty much unchanged by our knowledge of genetics. We still have the usual mix of kindness, friendship, nastiness, respect, disrespect, philosophy, and religion that exists everywhere else."

An Agenda for Scientists

Scientists face two main tasks in their attempts to reduce public misgivings about their work, according to Singer. The first is "to teach the substance of science more broadly and deeply." Widespread ignorance about science needs to be tackled starting with the youngest schoolchildren. For example, people's questions about the implications of mapping and sequencing the human genome will require scientists to assess the work's impact and answer the public's questions. Such educational efforts require time spent at meetings and hearings, talking to the media, and dealing with legislators, time that most scientists would probably prefer to spend on their research. But it is one of the only ways to counter the views of those who would see science constrained by playing on the public's fears of science.

The second task is "to convince nonscientists of the basically human nature of science and scientists." Scientists must reveal their own doubts and questions about the application of their work and be prepared to criticize "loud and clear" when the results of their work are misapplied, Singer believes. For instance, reports have surfaced that some parents are subjecting their children to unknown dangers by giving them human growth hormone, now plentiful thanks to recombinant DNA technology, to increase their stature for athletic or social reasons. "We should object," Singer says. "We should remind the public that evil deeds arise from human ignorance and greed, not from high-tech opportunities."

Scientists, and particularly biologists, should also seek to emphasize their concern that animals used in scientific research are treated humanely. More than anyone else, biologists recognize the value of animal research to basic biological knowledge and to human and animal health care. They are overwhelmingly opposed to those segments of the animal welfare and animal rights movements that would curtail this research. But scientists cannot let their opposition to these groups diminish their condemnation in those cases in which inhumane treatment does occur. "It will not hurt us—indeed, it will help public understanding of science—if we admit that such cruelty troubles us, too, or that some few members of our community may be willing to exceed easily recognizable norms and should be stopped."

Biologists must also be vigilant about the possible misuse of their work for biological warfare, Singer believes. The military contends that current work on biological warfare is purely for defensive reasons, "but research for defense and offense is not very different in this field," Singer says. "We have a special opportunity to align ourselves with human values.

Finally, scientists must guard against mythologizing their own work by looking to science for answers to every human problem. "In our fervor for science we too often forget humility," Singer says. "We forget that our ignorance far exceeds our knowledge."

2

Biological Development and Cancer

A sperm cell and an egg cell unite. This seemingly straightforward event sets in motion one of the most awe-inspiring processes in all of biology. Within a short time the fertilized egg begins to divide, producing two cells, four cells, eight. The dividing cells form a ball, which then becomes hollow. Parts of the ball form dimples and ridges, with layers of cells moving inside other layers. Soon these layers thicken, forming different kinds of tissues. Other cells leave their points of origin and migrate through the developing embryo, eventually to become vertebrae, muscles, nerves. Tissues fold and protrude to form organs and limbs. The outlines of a living creature take shape. And already within the growing embryo, the sex cells that someday will repeat the entire process are growing.

For centuries, biologists have been fascinated with the process by which a single fertilized egg gives rise to a complex organism. They have described the process in elaborate detail, developing a formidable vocabulary to label a growing embryo's stages and features. But biologists want to do more than describe the growth of an organism; they want to know how it occurs. What are the mechanisms responsible for the grand pageant of development?

Until recently, the vocabulary of developmental biologists "did not have much immediate context in terms of mechanisms," according to Marc Kirschner, professor of biochemistry at the University of California at San Francisco. But the rapid advances of molecular techniques have begun to bridge this gap. More and more, development is being explained in terms of the expression of genes and the corresponding molecular properties of cells and their environment. "We don't really expect that all our explanations of embryology are going to be in terms of linear sequences of nucleotides," says Kirschner. "But we want to be confident that these concepts could be reduced to that level."

Explaining the development of a fertilized egg in molecular terms is a daunting task, Kirschner concedes. Much of molecular biology has focused on purified components in well-defined environments. But developmental biology is much more complex. "What we're really lacking is a cell biology of multicellular populations," says Kirschner. "The progress in the recent past has been in the quite important area of single-cell biology. But if we're really going to understand development, we need to know how groups of cells behave in populations."

Unifying Concepts

Despite its overall complexity, developmental biology is built on a solid foundation of biological principles. Most important, it is clear that development is choreographed by the messages carried in an organism's genes. Nongenetic factors, including environmental factors, can also play a key role in development. But the genes establish development's basic pattern and timing. For this reason, a powerful form of experimentation in developmental biology is to look for genetic mutations that affect the way an organism develops.

Because virtually every cell in an organism contains the same collection of genes, an organism is essentially a clone of a fertilized egg. But the cells in that clone are radically different. These differences reflect the expression of different sets of genes within each cell. In other words, the cells have differentiated into nerve cells, muscle cells, skin cells, and so on (there are about 200 roughly distinguishable types of cells in the human body). In turn, the differentiation of a cell is the product of its history and its environment.

The difference between a human being and, say, a chimpanzee, is not so much the kinds of cells each has. Rather, it is the number of those cells and the way in which they are organized. During the development of an embryo, cells must reach their designated locations and assume their differentiated states. The central problem of developmental biology is to explain this process. How do cells know where to go? How do they know what to become?

The puzzle at the center of developmental biology can be stated another way, according to Kirschner. How does the tremendous complexity of an adult organism arise from the relative simplicity of a fertilized egg? Although an organism's genome contains large amounts of information, developmental biologists agree that development draws partly on information that is not located in an organism's DNA. This may seem to conflict with the concept that development occurs under the control of DNA. But development cannot be understood without recognizing other factors, even though these factors themselves may have genetic roots. "The real complexity that occurs in development comes after the egg stage," says Kirschner. "The egg provides a crude level of information, and very complex and exquisite mechanisms act to achieve higher levels of complexity."

Development occurs in all organisms that originate from the union of an egg cell and a sperm cell, and many processes in development are common to all such organisms. Religious and ethical constraints severely limit the research that can be done with human embryos (as described in the essay following this chapter). But developmental events that occur in other species are very similar to those that occur in humans, and often they can be studied more easily in other kinds of organisms. "Many of the same embryological problems that are confronted in vertebrate development are also confronted in invertebrate development," says Kirschner, "and some fundamental problems are confronted in all development—plants as well as animals."

Inside an Egg

The fruit fly *Drosophila melanogaster*, long a favorite of geneticists, has also shed a great deal of light on development. *Drosophila* is small and has a short life cycle (Figure 2-1), so that many organisms can be produced and studied in a short time. *Drosophila* also has a genome that is smaller than that of humans but still sophisticated enough to produce biological processes shared by many other organisms. (Each *Drosophila* cell has about one-twentieth as much DNA as a human cell, and its genome contains about one-tenth as many genes.) In addition, geneticists have developed powerful tools that have enabled them to manipulate the genome of *Drosophila* to a greater extent than with any other animal.

A major way of studying *Drosophila* is by screening adults or larvae for genetic mutations that affect some characteristic of the organism. Researchers can then work backward from the mutation to learn more about the genes responsible for the mutation. For example, one striking mutation, *Antennapedia*, causes *Drosophila* to grow a leg where an antenna should be (Figure 2-2). The gene responsible for this mutation

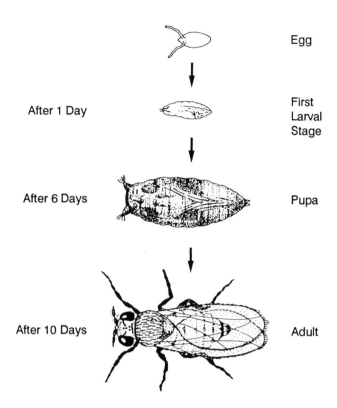

	Egg
After 1 Day	First Larval Stage
After 6 Days	Pupa
After 10 Days	Adult

FIGURE 2-1   The fruit fly *Drosophila melanogaster* develops from a fertilized egg into an adult in about 10 days. The egg forms an embryo that hatches into a larva in about a day. After several larval stages, *Drosophila* becomes a pupa and undergoes metamorphosis, giving rise to the adult fly. Reprinted, with permission, from G. Rubin, "*Drosophila melanogaster* as an Experimental Animal," Science 240(4858): 1443–1447, 1988. © 1988 by the American Association for the Advancement of Science.

FIGURE 2-2   The mutation *Antennapedia* causes *Drosophila* to grow a leg where it normally would grow an antenna. Mutations such as this offer valuable clues about the genetic controls over development. Reprinted, with permission, from B. Alberts et al., Molecular Biology of the Cell. New York: Garland Publishing, Inc., 1983. © 1983 by B. Alberts et al.

has been cloned and has been found to lie next to genes that control other aspects of *Drosophila*'s form. At this point, biologists have described and analyzed mutations in over 3,000 *Drosophila* genes (out of a total of 5,000 to 10,000 genes).

Fruit flies develop from eggs laid outside the mother's body. The early stages of development vary somewhat from the generic process described at the beginning of this chapter. Instead of undergoing successive divisions to produce a multicellular organism, the fertilized eggs of *Drosophila* take a shortcut (Figure 2-3). First, the DNA in the egg replicates, until the egg contains about 5,000 copies of its DNA. Then, about 3 hours after fertilization, cell walls form simultaneously around all of the nuclei in the egg, creating an embryo of about 5,000 cells. The embryo then undergoes a complicated set of invaginations before hatching as a larva.

In many ways, the *Drosophila* egg right after fertilization may seem to be a relatively simple, homogeneous structure (with the exception of the DNA in its nucleus). But careful observations reveal a richer texture. Soon after the dividing nuclei have migrated to the surface of the egg, they form 14 distinct bands along its length. Each of these bands will eventually develop into a specific part of the embryo, and ultimately the adult organism. Some bands will contribute to the head, others to the thorax, and others to the abdomen.

How do the nuclei in each band know what segment of the embryo to make? There are two limiting possibilities. One is that each band is influenced by a different molecular substance prelocalized to that portion of the egg. These substances could cause the DNA in specific

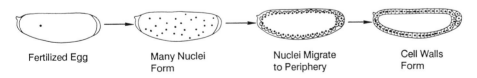

Fertilized Egg   Many Nuclei Form   Nuclei Migrate to Periphery   Cell Walls Form

FIGURE 2-3   After fertilization, the nucleus in a *Drosophila* egg divides many times, producing about 5,000 copies that eventually migrate to the periphery of the cell. Soon after cell walls form around the individual nuclei, the egg folds in upon itself to produce an embryo, which then hatches into a first-stage larva. Reprinted, with permission, from B. Alberts et al., Molecular Biology of the Cell. New York: Garland Publishing, Inc., 1983. © 1983 by B. Alberts et al. (After H. A. Schneiderman, pages 3–34 in Insect Development. Oxford, England: Blackwell, 1976.)

regions to express the genes appropriate for that position. The other possibility is that the dividing nuclei interact with some simpler spatial clue within the egg to produce the observed segmentation pattern.

Biologists have begun to solve this problem by examining mutations that distort the segmentation of *Drosophila*. For instance, some mutations cause larvae to develop missing segments, or double heads or tails, or other unusual features (almost all of these mutations are lethal, so that development stops at a preadult stage). Using this information, researchers have concluded that the spatial clues in the egg are relatively simple. In particular, it seems that the segmentation patterns observed in *Drosophila* are caused by substances emitted from either end of the egg. The concentrations of these substances drop as they get farther from their points of origin, so they form smoothly varying gradients across the length of the egg. Somehow the dividing nuclei read these gradients to determine where they are in the egg and what genes they should express. The proteins produced by these genes then establish additional gradients, which further elaborate the developmental pattern. "Rules are built into the genome and into the biochemistry of genes and their products that allow the expression of this very specific pattern of segmentally expressed genes," says Kirschner. "This presages the segmentation pattern seen later in development."

The segmentation of *Drosophila* demonstrates the complex way in which DNA interacts with its chemical environment during development, according to Kirschner. DNA does not just passively issue commands that are then carried out by the developing embryo. Rather, the expression of DNA is modified by its surroundings, first in the egg and later in the developing organism. The result is an intricate system of interrelated biological components that can produce complex developmental events. "I think this is the heart of the problem of embryology," says Kirschner, "how complexity is generated."

Segmentation is a basic feature of all advanced animals. In humans, segmentation can be seen in fingers and toes, arms and legs, backbones and ribs. By studying how segments develop in organisms like *Drosophila*, biologists hope to uncover mechanisms that govern similar processes in other organisms, including humans.

Communication Among Cells

Prelocalization of substances in the egg can be an important factor in early development (although in some animals, including humans, it seems to play a negligible role, as described in the box on pages 42–

43). But prelocalization is soon superseded by other factors. Most important, as cells divide, they begin to communicate among themselves. Without this communication, development would be impossible.

Many important aspects of cell–cell communication have been studied in the African clawed toad *Xenopus laevis*. As with *Drosophila*, the eggs of *Xenopus* develop outside the body, making the embryo comparatively easy to study. *Xenopus* eggs are also unusually large—about twice the size of the period at the end of this sentence.

Xenopus is also like *Drosophila* in that prelocalization of substances within the egg shapes early development. A *Xenopus* egg has a darker hemisphere, known as the animal pole, and a lighter hemisphere, known as the vegetal pole. As the fertilized egg begins to divide, this distinction between the two hemispheres remains.

A *Xenopus* egg undergoes a process of cell division and folding very similar to the one described at the beginning of this chapter. About eight hours after fertilization, the egg has divided into a hollow ball of about 4,000 cells. Shortly thereafter, a group of cells near the border of the animal and vegetal halves of the embryo move into the interior of the egg, where they come into contact with the animal pole of the embryo (Figure 2-4). These adjoining layers of cells then differentiate into the three kinds of tissue that will eventually give rise to all the parts of the adult body. The top layer of cells will form ectoderm, which will go on to form skin and the brain. The bottom layer of cells will form endoderm, which will produce the lining of the gut and associated organs, such as the lungs and liver. Between the two layers will form a third layer, the mesoderm, which will eventually generate muscle, ligaments, blood vessels, bones, the heart, and blood cells.

Kirschner points out that it is possible to draw what is known as a fate map of the early embryo, when it is still a ball of cells surrounding a hollow cavity (Figure 2-5). Cells derived from the animal pole of the embryo will make ectoderm. Cells derived from the vegetal pole of the embryo will make endoderm.

It is then possible to isolate various parts of the embryo and see how they develop on their own. When cultured in solution, cells from the animal part of the embryo form tissues characteristic of ectoderm. Cells from the vegetal cap form tissues characteristic of endoderm. Neither, however, when cultured on its own, can form tissues characteristic of mesoderm.

But if the two types of cells are placed in contact, the situation changes. Cells from the animal part of the embryo suddenly begin forming mesodermal tissues, such as muscle and kidney tissues. "These two crude regions, when mixed together in this manner, generate the complexity of the normal embryo," Kirschner points out. The implication is that some kind of chemical messenger travels from the vegetal cells to the animal cells and induces them to differentiate.

What is the nature of this messenger? It operates at very low concentrations, and it has been impossible to isolate and characterize directly from the embryo. So developmental biologists have tested other substances to see if they can cause cells from the animal portion of *Xenopus* embryos to form mesodermal tissues. "The problem," according to Kirschner, "was not so much the failure to find these substances, but the finding that lots of things would have effects." Substances like fish swim bladders and guinea pig bone marrow and even nonspecific

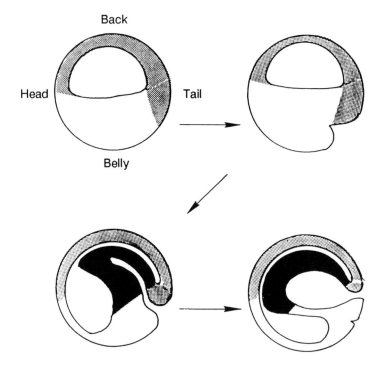

FIGURE 2-4 By the time the *Xenopus* egg has divided into several thousand cells, the embryo has formed a hollow ball. The cells on the top of the ball (gray) are derived from the animal pole of the egg; those on the bottom of the ball (white) are derived from the vegetal pole. During a process known as gastrulation, cells at the boundary of the two halves of the embryo move into its interior. These cells will go on to form endodermal tissues (black), while cells from the top of the embryo will generate ectodermal tissues. Mesodermal tissues will form in a layer intermediate between these two. Reprinted, with permission, from R. Dulbecco, The Design of Life. New Haven, Conn.: Yale University Press, 1987. © 1987 by Yale University.

factors like ammonia and calcium turned out to have a mesoderm-inducing effect. "This got to be very discouraging," Kirschner says.

More recently, Kirschner and his coworkers have been one of several research groups trying to find the substance that actually operates in the *Xenopus* embryo. First, they reasoned that the substance might be one of the known growth factors, which are proteins that bind to receptors on cells and cause them to divide. So Kirschner's team "wheeled our cart down the hall at UCSF and collected all the known growth factors to try them." The only one that had an effect was fibroblast growth factor, which performs a number of functions in the body, including causing blood vessels (a mesoderm-derived tissue) to proliferate. But even fibroblast growth factor did not produce as powerful an effect as the unknown substance operating in the embryo.

So Kirschner and his colleagues started combining growth factors. "You can spend your whole life doing this kind of thing," Kirschner says, but before long they found what they were looking for. When fibroblast growth factor was mixed with a particular kind of transforming growth factor, a protein that regulates cell differentiation, the combination caused induction at levels comparable to those found in the embryo. The transforming growth factor, known as TGF-beta, had no inducing effect by itself. But with fibroblast growth factor it acted synergistically, producing an effect much greater than fibroblast growth factor alone could produce.

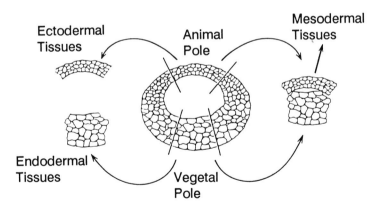

FIGURE 2-5 When cells from the animal or vegetal poles of the Xenopus embryo are cultured separately, they form tissues characteristic of ectoderm and endoderm, respectively (left). However, when cells from the two poles are cultured together, they produce mesodermal tissues (right). Reprinted, with permission, from I.B. Dawid and T.D. Sargent, "*Xenopus laevis* in Developmental and Molecular Biology," Science 240(4858): 1445. © 1976 by the American Association for the Advancement of Science.

"At this point we could justifiably ask whether we had gotten our hands on the most expensive, most purified nonspecific factor available," Kirschner points out. "Maybe this was just another form of fish swim bladder." But new findings indicated otherwise. Another researcher had cloned DNA copies of a messenger RNA that is prelocalized in the vegetal pole of *Xenopus* eggs. When the DNA was sequenced, it was found to be very similar to the gene that produces TGF-beta. In addition, Kirschner's team found that *Xenopus* eggs contain high concentrations of a protein very similar to fibroblast growth factor.

This evidence does not prove that fibroblast growth factor and TGF-beta together cause cells in the animal part of the *Xenopus* embryo to produce mesoderm, Kirschner cautions. "But it seems very likely that they're responsible," he contends. Work is continuing to identify the actual factors involved, and conclusive identifications are expected within a few years.

The induction of mesodermal tissues in *Xenopus* embryos is an example of a process that is critical in development. The chemical and

....

Is Information Prelocalized in Human Egg Cells?

Although prelocalization of substances in the egg shapes the early development of *Drosophila*, *Xenopus*, and many other species, it seems to play no part in the development of mammals, including humans. In mammals, the fertilized egg first cleaves into several dozen cells and then forms a hollow ball. A thickening of cells within one end of the ball, known as the inner cell mass, will eventually form the embryo. The outer sphere of cells, known as the trophoblast, will form the placenta.

Up to the eight-cell stage, the cells of a mammalian embryo appear to be identical. One way to demonstrate this is to separate a single cell from an eight-cell embryo and place it either inside or outside another eight-cell embryo. If inside the embryo, the introduced cell will form part of the fetus. If outside, it will form part of the placenta. This suggests that it is the position of cells either inside or outside the developing embryo, rather than any intrinsic feature of the cells, that determines whether they will become part of the fetus or the placenta.

Another way to demonstrate the similarity of cells in an early embryo is by separating them completely from one another and putting them back together in a different configuration. Despite this jumbling of position, the cells still produce a normal embryo.

A variant on these experiments demonstrates a remarkable capacity of early mammalian embryos. If two embryos at the eight-cell stage are pushed together under the proper conditions, they can fuse to form a single embryo (see figure). If implanted into a foster mother, such an embryo can develop into a so-called chimeric animal consisting of genetically different groups of cells. Such a chimera has four parents, with cells derived from each set of parents scattered throughout its body.

physical messages that a cell receives from its surroundings help it determine how to behave. It also demonstrates a common feature of chemical messengers. As with fibroblast growth factor and TGF-beta, the messengers active in development often function in the adult organism as well.

Development Gone Awry: Cancer

Development does not end after birth. Cells continue to grow, divide, and differentiate as the body grows and as old cells wear out. The attainment of sexual maturity during adolescence is a developmental process. So are many of the changes, such as milk production, that occur during pregnancy. Even aging and death can be thought of as a process of controlled development.

Some cells in the adult body, including nerve cells, the muscle cells of the heart, and the lens cells of the eye, cannot reproduce. Once

8-Cell Mouse Embryo Whose Parents Are White Mice

8-Cell Mouse Embryo Whose Parents Are Black Mice

Embryos Are Pushed Together and Fuse

Trophoblast — Inner Cell Mass

Embryo Transferred to Foster Mother

The Baby Mouse Has 4 Parents (But Its Foster Mother Is Not One of Them)

Two mouse embryos at the eight-cell stage can be fused to produce a chimeric mouse with four parents. The cells on the inside of each embryo form the inner cell mass, eventually to become the fetus, while those on the outside form the trophoblast, which becomes the placenta. Reprinted, with permission, from B. Alberts et al., Molecular Biology of the Cell. New York: Garland Publishing, Inc., 1983. © 1983 by B. Alberts et al.

established, they function until they die or the organism of which they are a part dies.

Other cells multiply by simple replication. For instance, when liver cells die, other liver cells divide to make up for the loss. The same process occurs in blood vessels, in the pancreas, and in other body tissues.

A third category of cells originate from undifferentiated cells known as stem cells. When a stem cell divides, each of its two daughter cells has two choices. It can remain a stem cell, or it can differentiate into another cell type. For instance, all blood cells—white cells as well as red cells—originate from a single type of stem cell.

According to Barry Pierce, professor of pathology at the University of Colorado, the unchecked proliferation of cancer cells bears an uncanny resemblance to the process of cell renewal. Cancer is a "caricature" of cell renewal, Pierce says, with "caricature meaning overproduction or gross exaggeration." In cancer, a stem cell or some other kind of cell capable of division is somehow transformed into a malignant stem cell, one that produces many cancerous offspring. Differentiation into other cell types may be blocked, with malignant stem cells producing more stem cells in an uncontrolled avalanche of proliferation.

Pierce has decided to take a developmental view of this process. "Cancer is a tissue," he says, "it's composed of cells. Does it obey the laws of developmental biology?" By viewing cancer in this light, it is possible to envision some striking alternatives to current cancer treatments.

One common misconception about cancer, Pierce notes, is that cancer cells always beget cancer cells. "This has had an unfortunate impact on developing alternatives to current therapies," he contends, "because it implied that unless cancer cells were eradicated from the body or destroyed, the host will die." Treatments have therefore focused on eliminating cancer cells entirely through surgery, chemotherapy, or radiation therapy. These procedures have met with some remarkable successes, Pierce points out, but they can exact a high toll on patients. Chemotherapy or radiation therapy do not just destroy cancer cells; they also destroy normal cells. "Oncologists must poison the patient and then rescue that individual," Pierce says. "Then they administer another dose of poison followed by rescue, with the hope that through repeated cycles the patient will recover and the tumor will suffer incremental damage, resulting in a cure." It is a "hair-raising" situation, Pierce observes, "and I think that doctors who do this are probably the best physicians in the world, because they walk such a narrow line, with death on each side."

But the dogma is wrong: cancer cells do not always give rise to cancer cells. There are a number of situations, Pierce notes, in which cancer cells differentiate into apparently normal cells. For instance, teratocarcinomas are cancers that arise from malignant sex cells. In the body, these cancers share the characteristics of both cancers and embryos. The tumors contain both undifferentiated cancerous stem cells and differentiated cell types found in embryos, including tissues derived from ectoderm, mesoderm, and endoderm.

The important feature of these tumors is that the differentiated cell types are derived by differentiation from the cancerous stem cells and are usually not cancerous. If extracted from the tumor and grown in culture, most of these differentiated tissues behave normally. In the

. . . .

Making Cancer Cells Into Normal Cells

One way to demonstrate that cancer cells do not always remain cancer cells is through an elegant experiment using mouse embryos. If a malignant sex cell known as a teratocarcinoma cell is grown in culture, it will typically go on to produce a tumorous mass. But if such a cell is placed into an early mouse embryo (see figure), it often will lose its cancerous characteristics, and the embryo will develop normally.

Even more striking is the fate of the injected cell. If the cell is implanted into the inner cell mass, it will become part of the developing fetus and eventually will contribute to tissues throughout the body. If the cell is implanted into the trophoblast, it will contribute to the placenta. Not only does the cell lose its malignancy, but it is regulated by the embryo in such a way that it acts as a normal embryonic cell. Even though such a cell may have chromosomal abnormalities, it responds to external stimuli in the proper way and can be considered normal.

Other kinds of cancerous cells, including leukemia and melanoma cells, can also lose their malignancy if inserted into specific parts of developing embryos at specific times. Despite being cancerous, the cells have the ability to respond to some factor produced in the embryo that can cause them to revert to normal cells and become a part of the growing embryo. It is at least possible that every kind of cancer cell could be regulated this way, although the mechanisms of this regulation remain unknown.

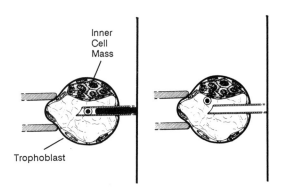

Cancer cells can be introduced into mouse embryos through extremely small injecting pipets. The larger tube (at left) is a pipet that holds the embryo in place through suction. Reprinted, with permission, from G. B. Pierce et al., Cancer Research 44:3987–3996, 1984. © 1984 by the American Association for Cancer Research.

process of differentiating, the cancer cells have somehow shed their malignancy.

There are other cases in which cancer cells have been shown to differentiate into nonmalignant cells. For instance, leukemia, colon cancer, and breast cancer cells can be chemically induced to differentiate into nonmalignant cells. Most remarkably, if certain kinds of cancer cells are inserted into embryos at particular times and places, they lose their malignancy and become part of the embryo (see box, page 45).

The ability of cancer cells to differentiate into normal cells points toward an entirely new approach to cancer therapy. If it were possible to induce cancer cells to differentiate, it might be possible to eliminate cancer cells without harming other cells in the body. These differentiation agents would have to be specific for the particular cancer cell, since agents that caused all normal stem cells to differentiate would be disastrous for an individual. The differentiated cancer cells must also remain benign, a condition that current examples do not always meet. But if these criteria can be met, differentiation therapy could offer a less toxic, more precise treatment for cancer than surgery, chemotherapy, or radiation therapy.

Pierce and other researchers are looking at two categories of potential differentiation agents. The first category consists of chemical substances similar to those used in chemotherapy. Researchers have shown that various chemicals can convert a variety of cancer cells to differentiated states. However, many of these substances are toxic, raising the same problems associated with chemotherapy.

The other category consists of substances that the body uses to signal cells to differentiate during development. For instance, TGF-beta, the transforming growth factor active in *Xenopus* embryos, also stabilizes the differentiation of particular cells in humans. Its effects on cancers have not yet been tested. But as biologists learn more about how cells are normally renewed in the body, they may be able to better control the diseases caused when cell renewal runs amok.

Ethical Considerations in the Use of Human Materials

*T*he first successful human transplant of a solid organ took place in 1954, when a young man in Boston, Massachusetts, received a kidney from his identical twin brother. Since then, scientific, technical, and medical advances have made organ transplantation a relatively common procedure. In 1987, some 1,500 Americans received transplanted hearts, approximately 1,200 received livers, another 1,200 received bone marrow, some 10,000 received kidneys, and approximately 35,000 received corneas. If not for a severe shortage of available organs, many thousands of other patients would receive transplants.

Biological materials taken from humans serve a number of other purposes besides transplantation, according to Arthur Caplan, professor of philosophy and surgery and director of the Center for Biomedical Ethics at the University of Minnesota. Human materials are essential in a wide range of biomedical research and training. Some human materials, such as blood plasma and sperm, are used for medical procedures unrelated to transplants. Human materials may even be used as exhibits for educational purposes. "One of my first memories of wondering about values in science occurred at the Harvard Museum, where I stood as a cub scout watching a display of fetuses in glass jars," Caplan recalls.

The sources of human materials are as diverse as their uses, spanning the entire range of human development. Human materials can come from sex cells, embryos, fetuses, abortuses, newborns, children, adults, and cadavers. In some cases, such as blood donations, human materials can be removed without harming the donor. In other cases, the materials are essential to life and therefore can be removed only after death.

Ethical questions have always surrounded the use of human materials, but rapid advances in transplantation procedures have brought these questions to center stage. How should a limited number of organs be allocated among a large number of potential recipients? What are the appropriate ways to obtain organs from potential donors? How should transplantation therapy be balanced against other forms of medical care?

More generally, bioethicists have asked: What human materials should ethically be used? Under what conditions should they be used? And what are the purposes for which they should be used? The diversity of sources and uses for human materials has generated a rich array of ethical questions, which will only become more numerous as medical capabilities increase.

The Inadequacy of Simple Rules

Much has been said and written about the ethical obligations of those who use human materials, but according to Caplan much of current practice can be boiled down to three simple rules. "One, don't worry about it unless the source of the material is a person. Two, treat everything with respect. Three, never violate either (a) God's will with respect to these materials or (b), if you don't believe in God, then nature's order."

The problem with these rules and other "simple nostrums" is that they leave crucial questions unanswered, Caplan says. In the case of the first rule, the obvious question is, What constitutes a person? One approach to this problem has been to try to equate the humanness or moral worth of an entity with its stage of development. For instance, an embryo can be seen as having more moral value than a sex cell, and a fetus as having more value than an embryo, and a newborn as having more value than a fetus. "When we think of the world around us, we tend to assign more and more moral standing to things the farther along they are in the hierarchy of development," Caplan points out.

Caplan believes that this approach is fundamentally wrong. The quest for some kind of moral essence that would make one entity human and another not is "a vain search," he says. Despite decades of effort, ethicists have not been able to find a distinctive property, such as consciousness or volition, that would allow them to make such a distinction. "Part of the emptiness of the search is proven by the fact that it has been going on for over 30 years," says Caplan. "Some very good people have sallied forth to try and find the magic property that confers moral worth on things, and they've come back not carrying anything."

The other two rules are equally problematic, according to Caplan. What does it mean to treat a human material with respect? Standards of respect vary, and it is difficult to derive a code of conduct from such a vague concept. As to the edict that humans should not intervene in the natural course of events, it is hard to know why not. Medicine is by its nature an intervention in natural processes. And it is difficult to be a living creature and not interfere to some extent with the natural course of events.

The Social Dimension of Ethics

If the users of human materials are to develop some sort of consistent framework within which to view their responsibilities, they must look beyond simple rules, Caplan believes. First, they must realize that moral worth is not intrinsic to any given object. Rather, it arises from the standing that society assigns to an object.

The social value of human materials derives partly from the wide-ranging implications of their use. For example, when a researcher uses a material to investigate a particular problem, the research does not just influence the researcher or the donor of the material. It potentially influences everyone, both through the research findings and through the precedent established for the future use of human materials. It is this social dimension that is missing in the search for moral essences, according to Caplan.

In examining the social dimensions of the issue, "certain values loom large as forming the moral environment in which we can interact with human things," Caplan observes. One of the most important values is that the user of these materials do no harm. As an extreme example, it is morally wrong to bring about the death of someone prematurely just to get materials from that person to benefit someone else.

Away from the extremes, the application of this standard can be more difficult. It implies, for instance, that creating human embryos simply for the purpose of doing research on them would be morally wrong. But sometimes human embryos are not created specifically for that purpose. *In vitro* fertilization programs, for instance, typically create more human embryos than are reimplanted into a prospective mother. As a result, spare embryos are potentially available for research. In this case, Caplan contends, it may not be morally wrong to conduct research on these embryos because they were not created with a harmful intent.

The edict to do no harm can be extended to produce a more positive statement. If faced with an inevitably tragic situation, the user of a human material should strive to redeem the situation by doing good. This is why it is acceptable to take advantage of a fatal accident to make human materials available.

"Notions of redemption and transformation are very critical in dealing with our bodies," says Caplan. Christian traditions speak of the reconstitution of the physical body into a spiritual body, and something quite similar happens when a physician takes an organ from an accident victim and transplants it into another person. In this way, an act that otherwise would border on the sacrilegious can be transformed into something laudatory. "It might be profane manipulating human materials to achieve certain goals or ends," Caplan says, "but it can somehow be made acceptable if there is the possibility of redeeming or transforming a situation that began as tragic."

Another factor in the moral framework is certainty, according to Caplan. When a person uses a human material that is essential to life, that person needs to know that the donor of the material is "at least doomed to die and, hopefully, dead." The desire for certainty has led to what Caplan calls "one of the strangest discussions of the twentieth century—an attempt to figure out what death is." In recent years, society has reached a consensus over the so-called brain death theory—that a person is no longer alive when the brain has irreversibly ceased all function. But this consensus disguises the questions that still remain about the nature of death. "Whatever death is, it probably doesn't have very much to do with the electrical measurements of the brain," Caplan asserts.

A third value that influences the use of human materials is the deeply felt need for permission to use a certain material in a certain way. This is part of the rationale behind donor cards. People in the United States are encouraged to fill out cards stating how they would like their bodies to be treated after death. Moreover, even if a person indicates that the transplant of organs would be acceptable, organs are generally not removed without first obtaining the permission of family members.

A final social value is related to the need for efficacy. The users of human materials must be able to demonstrate that they will be able to use those materials effectively.

The Origins of Social Values

An examination of these four values—do no harm (and if possible try to do good), be certain, try to get permission, and be effective—reveals something quite remarkable about their origins, Caplan points out. They are not the kind of values that would be imposed on researchers and clinicians by a society anxious to control the use of human materials. Rather, they are values that researchers and clinicians impose on themselves. "The drive to get ethical principles is not something that comes from without," says Caplan. "It's something that derives from within. It is very suggestive of what the researcher's or the clinician's values are about what ought to be done with human organs and tissues."

Caplan goes even further. The self-imposition of these values, he believes, answers a deep-seated need that the users of human materials have to be given permission for what they do. "It seems to me that 'respect' is a code word for something a bit different," Caplan says. "We want to make sure that those whose materials we use have given us exculpation to do it."

The Need for Equity

By recognizing the social origins of the values affecting the use of human materials, it is possible to draw a clearer picture of the moral obligations incumbent upon the users of those materials. For example, a key consideration in the use of human materials is that they be distributed equitably and fairly. If the public perceives inequities in the distribution of organs acquired through public appeals, they will come to have a "jaundiced view" of organ donation, according to Caplan. "When people believe that the rich get access to transplants but the poor don't, or that publicity plays a role in who lives and dies, it has a decimating impact on people's willingness to put faith or trust or confidence in those who would deal with biological materials."

One concrete implication of the need for equity, Caplan points out, is that the users of human materials should conduct much more follow-up than is typically done with the families of organ donors. Often, after families grant permission to use the organs of a relative, they are kept unaware of how those organs are used. By letting families know how the organs of a relative contributed to another person's life, misgivings about the distribution of organs will diminish.

The need for equity also demonstrates why many people object to putting monetary values on organs and distributing them through the marketplace. Such a practice can be seen as unfairly affecting the distribution of human materials. "Money is not in itself evil or dirty," Caplan says. "But what bothers people is the perception that money or commercialization will adversely affect the distribution of materials."

Public perceptions about the distribution of organs and tissues will play a large part in determining how they will be used in the future, Caplan believes. "Concerns about fairness, about equity, about how justly we distribute what it is we can do with respect to human biological materials is going to demonstrate society's willingness to tolerate their use."

3

Neuroscience and Neuronal Replacement

The crowning triumph in human development is the growth of the brain. By the time of birth, the brain contains several hundred billion nerve cells. An average nerve cell makes connections with thousands of other nerve cells, meaning that there are literally trillions of neural connections in the brain. It is nature's most astonishing and daunting creation, an organ that can hope to understand itself.

No new neurons are generated in the human brain after birth (although recent research has revealed a remarkable capacity in other species for the generation of new neurons, as described later in this chapter). But a newborn's brain cells are hardly immutable. They grow and make new connections; existing connections become stronger or weaker; droves of brain cells die, especially during the first few years of life, as if a block of marble were being sculpted into a statue. In this way the brain refines its abilities to direct movement, speech, memory, perception, emotion, reason—all of the vital activities encompassed by the rather nondescript word *behavior*.

The ultimate goal of neuroscience is to explain in biological terms the mechanisms of behavior. It seeks to answer such questions as, How do we remember what we have learned? How do we understand written or spoken words? How do we perform skilled movements? These questions can be studied at many levels. At the microscopic level, neuroscientists explore the properties of nerve cells and their interactions (see box, pages 56–57). At the macroscopic level, they examine the functions of specific parts of the brain and how those parts work together to produce a specific behavior.

Since shortly after Darwin's time, a prominent view of the function and structure of the brain has been based on the theory of evolution. According to this view, the brain evolved by adding or expanding sections on top of previous sections, culminating in the human brain (Figure 3-1). Each new section added new and more complex behaviors while suppressing the more primitive behaviors generated by lower brain regions.

This evolutionary perspective has produced valuable insights, according to Vernon Mountcastle, professor of neuroscience at the Johns Hopkins University School of Medicine, but in the last few decades it has been superseded by a more complex view. Today, the higher functions of the brain are seen not as the product of any one level or part of the brain but as the result of simultaneous activity in a number of interconnected regions. Neuroanatomists have shown that the intercon-

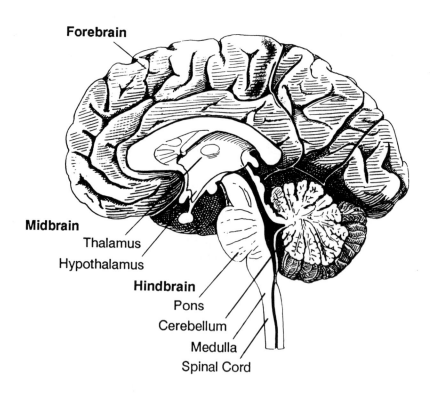

FIGURE 3-1 A cross section through the center of the human brain shows its major anatomical divisions. The hindbrain, midbrain (which has few distinctive features in humans), and forebrain are common to all vertebrates, but the characteristics of these sections vary greatly from species to species. Reprinted, with permission, from F. E. Bloom et al., Brain, Mind, and Behavior. New York: W. H. Freeman and Company, 1985. © 1985 by Educational Broadcasting Corporation.

nections between different parts of the brain are far more numerous than previously realized. The resulting integration of the brain's function argues more for a distributed than a hierarchical view of how the brain works. For instance, Mountcastle points out, motions initiated by the cerebral cortex may be modified and refined by the cerebellum. "Can you say that the cerebellum, a very ancient part of the brain, is lower than the supplementary motor area, a relatively new area? No. It doesn't make sense to say that."

High-Level Versus Intermediate-Level Functions

The greatest single challenge in neuroscience is to understand the higher functions of the human brain—thought, memory, emotion. Scientists have made great progress over the years in learning about these functions. They have pinpointed parts of the brain involved in such functions, and they have studied analogous activities in nonhuman species. But most of the important features of the higher functions produced by the human brain remain mysterious. They seem to involve many neurons spread throughout the brain interacting in ways that have not yet been described.

Nevertheless, neuroscientists are now poised to make rapid progress in answering what Mountcastle calls "intermediate-level questions," such as how the brain perceives its surroundings, how it attends to one sense to the exclusion of others, and how it modifies behavior in the light of past experience. In humans, these functions tend to involve the outermost portion of the brain, known as the cerebral cortex (although, as noted earlier, other brain regions also play roles). The uppermost portion of the cerebral cortex, also called the neocortex because of its relatively late appearance in evolution, consists of a quarter-inch-thick layer of neurons that, if laid out as a flat sheet, would cover a total area of about a foot and a half. This neural mass, deeply wrinkled and split into two lobes, blankets the top of the brain and gives the human brain its characteristic appearance. It is the greatly increased size of the neocortex that distinguishes primates from other vertebrates and humans from other primates.

The activities of the cerebral cortex that are best understood are the ones closest to the input and output functions of the brain. With regard to input, the cerebral cortex receives information from sensory detectors of the body located in the eyes, skin, ears, nose, and mouth. These receptors convert environmental stimuli into nerve impulses, which then travel through various way stations to the cerebral cortex and other brain regions. There, through mechanisms that are still poorly understood, these impulses undergo transformations leading to perception and other higher functions. If these stimuli provoke a physical response, the motor neurons of the neocortex come into play. They generate impulses, drawing upon input from other parts of the brain, and these impulses are in turn modified before reaching the muscles and causing movement.

Figure 3-2 shows the parts of the neocortex that receive information most directly from the senses (the primary sensory areas) and that initiate

....

The Building Blocks of the Brain

"The highest activities of consciousness have their origins in physical occurrences of the brain just as the loveliest melodies are not too sublime to be expressed by notes."
—W. Somerset Maugham
A Writer's Notebook (1896)

Neuroscience is based on the assumption that human behavior arises from the actions of electrically charged ions shuttling back and forth through the surfaces of nervous system cells. These cells, known as neurons, occur in a bewildering variety of shapes and sizes (see figure). But most of them share several common features.

Neurons communicate with each other by sending signals through long protrusions that extend from their cell bodies. One of these protrusions is an axon, which carries signals away from the neuron toward other cells. An axon may connect with just one other cell, or it may branch toward its tip and connect with many other cells.

The other extensions on a neuron are dendrites. Along with the cell body, dendrites receive messages from the axons that terminate upon them. Thousands of axons may converge upon a single neuron, and it is the dendrites and cell body that integrate the incoming messages and determine the cell's response.

Neurons maintain an electrical potential across their surfaces by pumping electrically charged atoms across their cell membranes. When a given neuron receives enough stimulation from other neurons, the electrical potential at the base of its axon abruptly reverses. This sends a wave of reversed electrical potential rushing down the axon. This signal, which is known as an action potential, always has the same magnitude and duration; it is an all-or-nothing proposition. For a neuron to send a stronger message, it therefore has to generate more action potentials in a given time, as if it were being forced to communicate in Morse code using only dots.

Electrical charge does not just jump across the junctions between neurons, like a spark of static electricity between a hand and a doorknob. It takes a more indirect but ultimately much more flexible route. When an action potential reaches the terminals of an axon, it triggers the release of certain chemicals into a gap, known as a synapse, between the sending and receiving cells. These neurotransmitters influence the electrical activity of the receiving cell, either encouraging it to fire or inhibiting its firing. (They can also modify the cell in longer-lasting ways, perhaps acting as a basis for some kinds of memory.) Over 50 different neurotransmitters have been discovered, with many more probably remaining to be found.

movement (the motor area). The interesting thing about Figure 3-2 is the large expanse of the neocortex that is not directly connected to the outside world. Neuroscientists have traditionally referred to these regions as association areas. Here, they hypothesize, information from the more specialized parts of the brain is compared and integrated. This information may include not only sensory input but emotions, memories, motivational states, and other factors. The comparison of sensory input with internally generated information is a key step in perception, the conscious or preconscious awareness of objects and

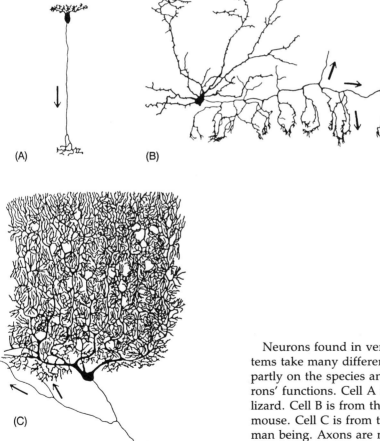

(A)

(B)

(C)

Neurons found in vertebrate nervous systems take many different forms, depending partly on the species and partly on the neurons' functions. Cell A is from the retina of a lizard. Cell B is from the cerebellum of a mouse. Cell C is from the cerebellum of a human being. Axons are marked with arrows (the full extent of the axon is not shown for C), and the drawings are not to scale.

actions in the surrounding world, as well as in other higher functions.

It has been very difficult to demonstrate the integrative activity of the association areas experimentally. But researchers are now beginning to explore these areas in meaningful ways, and in doing so they are beginning to understand the mechanisms responsible for many of the brain's remarkable abilities.

The Visual World

The route from sensory input to perception is better understood for vision than for any other sense. Everything a person sees in the visual field is refracted through the cornea and lens to form a small, inverted image on the back of the eye (Figure 3-3). There, the light-sensitive rods and cones generate electrical impulses corresponding to the light falling on them. These electrical impulses pass through several layers

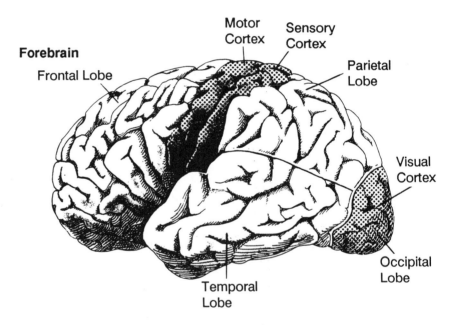

FIGURE 3-2 The sensory, visual, and auditory cortexes (the last is tucked inside the temporal lobe) receive impressions directly from the outside world, while the motor cortex controls complex movements. According to theory, the other parts of the neocortex, known as association areas, integrate impressions of the external world with internal information, a key step in such higher functions as the perception of surrounding objects. Reprinted, with permission, from F. E. Bloom et al., Brain, Mind, and Behavior. New York: W. H. Freeman and Company, 1985. © 1985 by Educational Broadcasting Corporation.

of neurons before exiting the eye through the optic nerve. Thus, the neural representation of the visual field has already undergone some initial processing before it reaches the brain.

After passing through the optic chasm, which directs everything from the right side of the visual field to the left part of the brain and vice versa, the optic tract sends most of its nerve fibers to a pair of structures in the thalamus known as the lateral geniculate nuclei. (These structures got their name because they resemble bent, or genuflected, knees.) From this way station in the brain, axons carry the visual signals to the primary visual cortex at the very back of the cerebral cortex. At the same time, some of the fibers of the optic tract go to three subcortical visual centers, which are involved in the control of eye movements, the regulation of light-induced hormonal changes, and various visual reflexes.

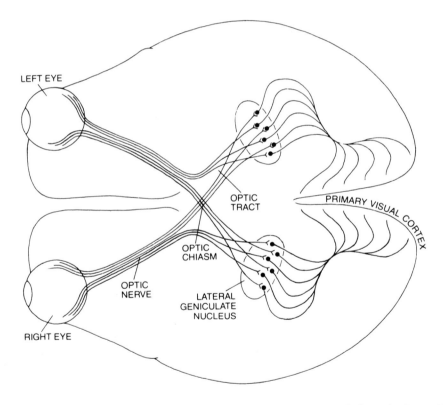

FIGURE 3-3 Neural impulses generated by the retina travel through the optic nerve to the lateral geniculate nuclei in the thalamus. From the thalamus, visual information travels to processing centers in the primary visual cortex and other brain regions. Reprinted, with permission, from D. H. Hubel and T. N. Wiesel, "Brain Mechanisms of Vision." Pages 85–96 in The Brain. New York: W. H. Freeman and Company, 1979. © 1979 by Scientific American, Inc.

The primary visual cortex is the first of the brain's processing centers for visual information. Neuroscientists have examined its function by inserting microelectrodes into the brains of experimental animals, primarily cats and monkeys. These microelectrodes record the electrical impulses generated by single neurons. When a specific visual stimuli causes a nerve cell to fire, the microelectrode can record the electrical impulses generated. In this way, researchers can determine what kinds of stimuli maximally stimulate a given cell, which in turn indicates the neural processing that the signals generated by the stimuli have undergone.

The primary visual cortex performs very specific operations on the nerve impulses that reach it from the lateral geniculate nuclei. Some neurons in this region respond only to bars of light oriented in particular directions. Others require that the bars be moving in a given direction at a certain speed. Others fire when presented with a boundary between light and dark regions. In this way, the primary visual cortex analyzes a complex scene by extracting information about the orientation and movement of the boundaries making up that scene.

If this were the only visual processing that occurred in our brains, the world would be an odd-looking place. But the primary visual cortex is only the first of many brain regions that analyze the visual world. Neuroscientists have found as many as 20 different areas in the cerebral cortex of monkeys that receive projections directly or indirectly from the primary visual cortex. Though many of these regions are still poorly understood, each of them seems to extract additional information from the nerve impulses coming from the eyes. For instance, one may be involved in the analysis of color, another texture, another form, another movement.

Vision in an Association Area

Mountcastle and his coworkers have been studying the neurons in one of these visual processing areas—the posterior parietal cortex, which is right behind the part of the cerebral cortex responsible for initiating movement (Figure 3-4). The posterior parietal cortex is one of the association areas of the brain, and studies suggest that it does in fact serve an integrative function. It receives information from the eyes, from the muscles and joints, and from the parts of the brain involved in attention. Studies of humans who have suffered an injury to the parietal lobe indicate that one of its major roles is to maintain

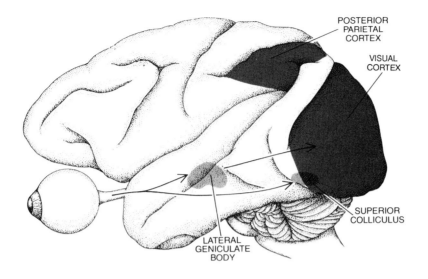

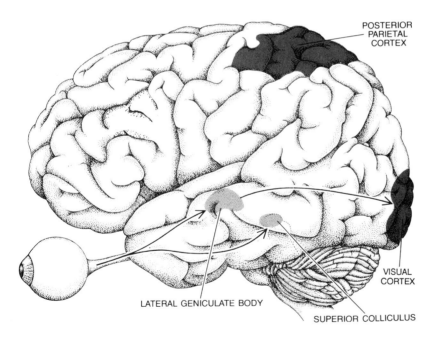

FIGURE 3-4   The primary visual cortex of the macaque monkey (top) is relatively much larger than that of humans (bottom). (Though shown here as the same size, the macaque's brain is actually less than half the size of a human's brain.) The posterior parietal cortex is located in comparable areas of the two brains and seems to serve very similar functions. Reprinted, with permission, from R. H. Wurtz et al., Scientific American 246:126, 1982. © 1982 by Scientific American, Inc.

an internal sense of the body's relation to the immediate environment (see box, below).

Only about 30 percent of the neurons studied by Mountcastle in the posterior parietal cortex respond to visual stimuli. Another 15 percent react not to visual stimuli themselves but to objects that are rewarding (such as food), novel, or aversive. About 10 percent of the neurons in this area are active before, during, or after certain types of eye movement. Another 15 percent are active during the process of reaching for something. And the function of the rest—about 30 percent—could not be ascertained by any of the tests that Mountcastle and his colleagues could devise.

Mountcastle's experiments were done using microelectrodes in macaque (rhesus) monkeys. In the 1960s it became possible to do such

....

Parietal Lobe Syndrome

Many insights into the function of the brain have come from studying the behavior of people whose brains have been partially damaged through stroke, injury, or disease. One of the strangest of these conditions involves damage to one of the parietal lobes and is known as parietal lobe syndrome.

Patients with the syndrome can still see and move about. But they tend to lose all interest in everything on one side of their bodies. One patient complained that someone had put a foreign arm in bed with him, even though he could feel normally with his arm. Others may forget to put on one of their shoes, or may shave just one side of their faces, or may eat from only one side of a plate.

These patients also have defects of visual and spatial perception (see figure). They cannot perceive their relationship to objects around them. Nor can they reach out to touch objects as easily as they could before.

Neurons cannot be regenerated in humans once they are destroyed. But sometimes other parts of the brain can take over

at least some of the functions of the regions that have been damaged. Thus, the bizarre symptoms of parietal lobe syndrome can fade over time as undamaged parts of the brain learn to process sensory inputs from the neglected side of the body.

A person with parietal lobe syndrome who was asked to draw a clock face put all of the numbers on the right-hand side, reflecting the brain's lack of attention to events occurring on the left-hand side of the body.

experiments in awake animals performing specific tasks, whereas before the animals had to be anesthetized; since then, this technique has become one of the most valuable in neuroscience. Essentially, a portion of the monkey's skull is replaced during surgery with a sterile metal plate. Microelectrodes can then be passed into the monkey's brain through the plate without causing the animal any pain, since the brain does not contain pain receptors. Meanwhile, the monkey can be trained to perform certain tasks. Once the monkey has learned those tasks, researchers can sequentially monitor the activities of hundreds of different neurons over periods of weeks or months.

In Mountcastle's experiments, monkeys were trained to fix their eyes upon a dim red light projected on a screen directly in front of them. When the light dimmed further, the monkeys pressed a bar and were rewarded with a drink of water. To make sure that the monkeys kept looking straight ahead, a computer monitored the position of their eyes and terminated the experiment if their gaze shifted. Meanwhile, a square block of light was projected elsewhere on the screen. The monkeys did not look at the square of light, but the microelectrodes indicated that information about it was being sent to neurons in the monkey's posterior parietal lobe.

One of the first things that Mountcastle's team observed was that the neurons in the posterior parietal lobe that reacted to visual stimuli (known as parietal visual neurons, or PVNs) had very large receptive fields. In the primary visual cortex, neurons respond to light only in a small area of the visual field. This is because of their close connections with the photoreceptors in the eye, which register the light from just a small portion of the image refracted onto the retina. But individual PVNs fired whenever the square of light appeared in large areas of the screen. Many even fired when the stimulus was on either side of the monkey's point of fixation, indicating that, unlike neurons in the primary visual cortex, individual PVNs receive input from both eyes.

The PVNs also exhibited a characteristic that Mountcastle calls foveal sparing. When the square of light approached the point of fixation, many of the PVNs stopped firing. This is the exact opposite of what happens in the primary visual cortex. There, a disproportionate number of neurons respond to objects directly in front of the eyes. This enables us to see the things that we are looking at more clearly than things that are in the periphery of our vision. But the PVNs behave differently; they respond best to objects that are in the periphery of the visual field.

Most of the PVNs studied by Mountcastle were also very sensitive to the movement of the square of light. The speed of the movement was relatively unimportant. But some neurons would only fire when the

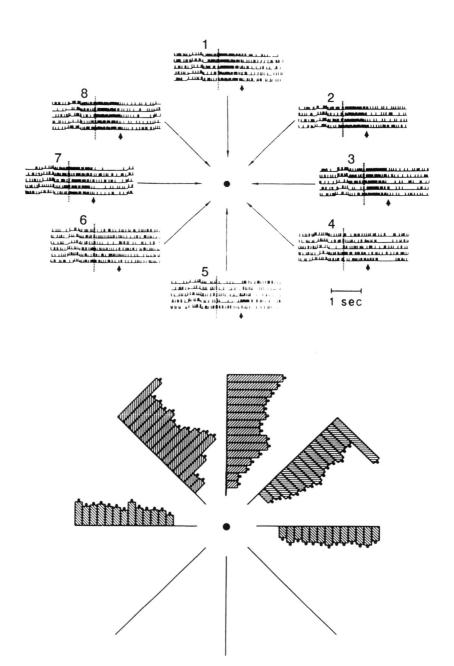

FIGURE 3-5 The response of a single visual neuron in the posterior parietal lobe can be measured by a microelectrode as a square of light moves in different parts of the visual field. Each vertical stroke in the readings in the top half of the diagram indicates a single discharge of the neuron. The experimental animal kept its gaze fixed on the dot in the center of the visual field. A light then appeared in the periphery of the visual field moving toward the point of fixation in the directions shown by the long arrows. The appearance of the light is indicated in the readings by the dotted

square was moving, and others could detect a moving stimulus in parts of the visual field where a stationary stimulus had no effect.

Furthermore, almost all of the neurons were sensitive to the direction in which the stimulus moved. Some fired only when the square moved toward the point of fixation; others fired only when it moved away. What's more, most neurons exhibited a distinct preference for one direction over others. For instance, one might fire only when the stimulus was moving away from the point of fixation toward the upper left at a 45 degree angle to the horizontal. Another might fire only when the block of light approached the fixation point from below.

Most of the neurons studied were not this specific. The neuron described in Figure 3-5, for example, fired whenever the block of light was moving toward the point of fixation in the upper half of the visual field. But almost every neuron had a "best" direction, defined as the direction of movement in which it would exhibit its maximum response.

Extending this analysis, Mountcastle and his colleagues showed how this "best" response could give a very precise indication of the direction in which something is moving in the peripheral visual field. Because most PVNs, such as the one in Figure 3-5, respond to movement in several different directions, a single neuron cannot indicate an exact direction of movement. Instead, the brain may rely on the responses of a large number of neurons to extract this information. By taking the "best" direction of a PVN and assigning that direction a magnitude based on the intensity with which that neuron responds to movement along any given direction, a vector analysis can be made of a population of PVNs (Figure 3-6). Viewed in this fashion, the total signal from the PVNs reflects the actual movement of the stimulus quite closely.

No mechanisms have yet been described in the brain that could perform this kind of vector analysis. But it could be a common way in which the brain circumvents the imprecision of individual neurons in making fine discriminations.

---

line. The time at which it crossed the center point is indicated by the small arrows. The procedure was repeated for the neuron several times for each of the eight directions of movement shown. Before the stimulus appeared, the neuron fired at a resting state. Movement of the stimulus toward the fixation point anywhere in the top half of the visual field greatly increased the discharge rate. A bar graph (bottom) shows the statistical increase in the discharge rate over the resting state. Reprinted, with permission, from M. A. Steinmetz et al., The Journal of Neuroscience 7:177–191, 1987. © 1987 by the Society for Neuroscience.

A Sense of Where You Are

Although the experimental conditions used by Mountcastle are quite abstract, the functions of the PVNs give a good indication of their roles in everyday activities. "It is the visual flow through the periphery that we use to control posture and locomotion, drive automobiles, and land airplanes," Mountcastle says. "We're able to compute the relative velocities of objects moving across the periphery of our visual fields with great accuracy."

If we are walking forward and looking straight ahead, the object we are looking at does not move. But everything else in our visual field seems to be moving. We do not consciously monitor this movement,

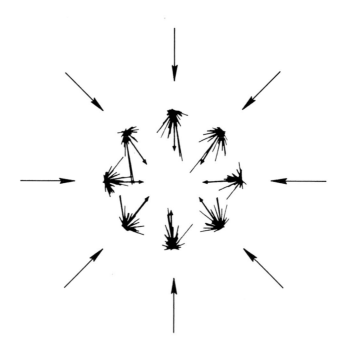

FIGURE 3-6 A vector analysis for eight directions of inward movement shows how populations of parietal visual neurons can provide a precise measure of the direction of movement. The small line segments for each of the eight directions correspond to the "best" direction of an individual parietal visual neuron. The length of the line segments corresponds to the relative rate of discharge when the stimulus is moving in the direction indicated by the outside arrows. When the line segments are added as vectors, the arrows extending from their common origin result. Note that this population vector differs very little from the actual direction of movement. Reprinted, with permission, from M. A. Steinmetz et al., The Journal of Neuroscience 7:177–191, 1987. © 1987 by the Society for Neuroscience.

but the PVNs receive input about it. For instance, if a runner approaches a branch extending into a path, PVNs seem to be involved in calculating the location of that object in space, allowing the runner to avoid it without ever looking directly at it.

In fact, the brain seems to have two separate systems for monitoring objects in the environment—one for objects in the center of the visual field, and another for objects in the periphery. The former system is one we might use to put on a ring; the latter to turn the page of a book. These two systems have different properties and seem to draw partially on different parts of the brain. For instance, it is very difficult to move your hand accurately while not looking at it when it is illuminated by a strobe light. But if you look directly at your hand, precise movements are possible even under strobe illumination.

The ability to detect the spatial arrangement of objects in the peripheral visual field and control movements of the body in the periphery have been very useful to humans, Mountcastle observes. When using tools or weapons, for instance, many movements have to be made in the periphery while attention is directed elsewhere.

Another interesting feature of PVNs is their dependence on levels of attention. When the macaques in Mountcastle's experiments were actively fixating on the dim red light in front of them, their PVNs fired an average of several times more strongly in response to the square of light than when they were just looking at the center of the screen. In other words, the more attention the monkeys focused on the task at hand, the more active the PVNs were in monitoring the peripheral visual field.

This, too, runs counter to common experience. When a person is focusing on a difficult task, less conscious attention is paid to objects in the surroundings. But preconsciously the PVNs have stepped up their activity. If a person senses something out of the corner of an eye, he or she can immediately bring that preconscious perception to full consciousness by looking at whatever moved. The adaptive value of such a system to early humans, who must often have had to concentrate on manual tasks while remaining wary about enemies, is obvious, Mountcastle observes.

Representations of the World

In broad terms, the activities of the PVNs are comparable to the activities of other light-sensitive neurons throughout the brain. Essentially, they receive nerve impulses directly or indirectly from the eye and transform those impulses into a more complex representation of the visual world. It is the same process that neurons in the primary visual cortex undertake in analyzing the orientation and movement of boundaries, or that neurons in other brain regions undertake in analyzing color or depth or texture.

The question still remains as to what part the PVNs play in the overall process of perception. Brain scientists have often wondered if all the visual representations generated by different parts of the brain eventually travel to a single destination, where they are combined into a unified perception. Often they phrase this question in a more evocative way: Is there a single cell somewhere in the brain that fires only at the sight of one's grandmother?

Mountcastle believes not. He believes that perception and other higher mental functions are distributed processes that draw both sequentially and in parallel on many different parts of the brain. "It is a succession of neural transformations," he says, "a flow-through from sensory input to motor output." The actions of the PVNs are therefore one part of this distributed system. When these neurons are damaged, as in parietal lobe syndrome, the process of perception is correspondingly degraded, but it is not destroyed.

The distributed nature of perception is mirrored in the importance of populations of neurons in the posterior parietal cortex. "It's highly unlikely that we will be able to understand the neurological higher functions if we understand everything about the function of single cells," says Mountcastle. "Reductionism is not enough."

To this end, Mountcastle has begun conducting experiments using arrays of microelectrodes that can detect the activity of several neurons simultaneously. With this technique he hopes to explore the dynamic interactions of groups of neurons working together, rather than single neurons working in isolation. "As we approach the study of higher functions," he says, "we have to study population activity."

Neural Regeneration in Canaries

Finding the exact sources of higher functions in the human brain has been an elusive goal. But in other species it has been possible to track down the brain areas responsible for quite complex behaviors. A prime example comes from studies of canaries carried out by Fernando Nottebohm, professor of zoology at Rockefeller University, and his coworkers. Their research has also overturned a long-standing belief in neurology: that the generation of new neurons is rare or nonexistent in the brains of grown vertebrates.

Baby canaries typically hatch in the spring. After a month or so, the males begin to sing, first in a soft, unformed voice known as subsong and then in a stronger but still immature voice known as plastic song. "It almost seems as if the animal were playing with his vocal behavior without having a message to convey, just trying out the properties of his vocal tract," says Nottebohm.

For 6 to 10 months canaries imitate the songs of other canaries, gradually building up a repertoire of distinctive syllables or phrases. Researchers can record and identify these songs on a spectrograph, a device that converts sound waves to traces on graph paper. Finally, by late fall or winter, the canaries have progressed to full song, which typically consists of several dozen distinctive syllables.

The spring after they are born, male canaries begin to breed with females, using the songs they have developed to attract mates and mark their territories. But their songs do not last. As soon as the breeding season is over, the males' songs become unstable, reverting to the plastic song that had characterized their singing a year earlier. During this period, the males learn new syllables, forget some, and modify others. Then, in the late fall and winter, full song returns, to be used in the next breeding season. Throughout the canary's lifetime of 10 years or so the cycle repeats itself—full song during the breeding season followed by plastic song.

Canaries sing with an organ in their throats known as a syrinx. The centers in the canary brain that control the syrinx begin with the hyperstriatum ventralis, pars caudalis, abbreviated HVc (Figure 3-7). The HVc, which is situated right beneath one of the fluid-filled ventricles of the canary brain, receives projections from the brain's auditory region, indicating that it is at least partially involved in interpreting songs the canary hears. The HVc, in turn, sends messages to a clump of nerve cells known as the robust nucleus of the archistriatum (RA). The RA projects to a nucleus in the hindbrain of the canary, which contains the motor neurons that send their axons to the syrinx.

The syrinx is an unusual organ in that it is involved only in singing. As a result, points out Nottebohm, the brain centers responsible for the canary's singing can be identified quite clearly. "It is very seldom that you can point to a part of the brain and say that this part is involved in learning to play the piano or speaking and nothing else."

The boundaries of the HVc and RA can also be defined quite precisely, because of the distinctive appearance of the neurons they contain. By measuring the size of these brain regions, Nottebohm and his team have been able to correlate the volume of these regions with the singing behavior of individual birds. They have found that canaries with large repertoires of song syllables have larger HVc's and RA's, while birds with relatively small HVc's and RA's tend to have smaller repertoires.

The HVc is also four times larger in adult males than in adult females, and the RA is three times larger. This makes sense, Nottebohm says, since male canaries sing and females do not. But this sexual difference is not inviolate. If given the hormone testosterone for a few weeks, females also begin to sing, though with a more limited repertoire than males have. At the same time, the volume of their HVc's nearly doubles, and their RA's grow by about 50 percent.

The most remarkable discovery made by Nottebohm and his colleagues is that the volume of the HVc and RA in male adult canaries fluctuates greatly over the course of the year (Figure 3-8). Right after the breeding season, when full song reverts to plastic song, the size of the HVc drops by nearly half. At the same time, the levels of testosterone generated by the male plummet, indicating a close correlation between hormone levels and the sizes of these brain regions.

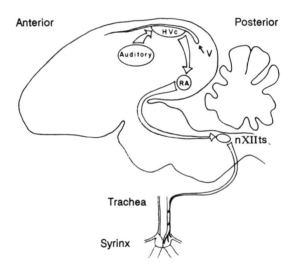

FIGURE 3-7 The hyperstriatum ventralis, pars caudalis (HVc) of the canary brain, which is situated directly beneath the lateral ventricle of the forebrain (V), receives input from the auditory regions of the brain and projects to the robust nucleus of the archistriatum (RA). The RA sends axons to part of the hypoglossal nucleus (nXIIts), which innervates the muscles of the syrinx, the organ canaries use to sing. Reprinted, with permission, from F. Nottebohm, The Condor 86:227–236, 1984. © 1984 by the Cooper Ornithological Society.

New Neurons or Old?

There are two possible ways in which the volumes of the HVc and the RA could change so dramatically. The first would be if existing cells in these areas grew, say by forming new dendrites and synapses. Indeed, this appears to account for at least part of the increased volume, Nottebohm observes. When females are given testosterone, the dendrites in the RA's of the females grew by an average of 50 percent, and the number of synapses in the RA increased by 70 percent.

The other possibility is that new neurons are being generated and incorporated into the brain, in direct violation of the dogma that no new neurons are generated in the adult brains of vertebrates. To test this theory, Nottebohm and his coworkers injected canaries with thymidine containing radioactive hydrogen. Thymidine is one of the four bases incorporated into DNA, so when a cell getting ready to divide reproduces its DNA, it will use the radioactive thymidine to form DNA.

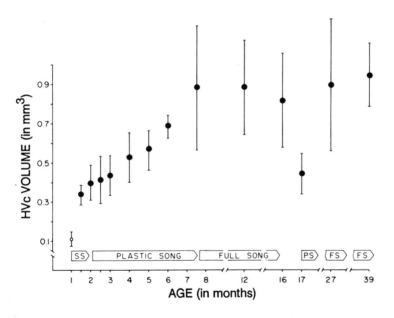

FIGURE 3-8 The volume of the hyperstriatum ventralis, pars caudalis (HVc) varies dramatically over the course of a male canary's life. It reaches its greatest extent at the onset of full song in the winter and spring and drops by about half after every breeding season. The vertical bars indicate one standard deviation of the data from groups of canaries. Subsong (SS) is a young canary's unformed sounds, plastic song (PS) is a stronger but still immature song, and full song (FS) is the adult song used during the breeding season. Reprinted, with permission, from F. Nottebohm, The Condor 86:227–236, 1984. © 1984 by the Cooper Ornithological Society.

In this way, the daughter cells become radioactive, and new cells can be identified by the radiation they give off.

When Nottebohm and his colleagues performed this experiment on male and female canaries, the results were hard to believe. They found that new neurons were indeed being generated, not only in the HVc but throughout the birds' forebrains. Furthermore, they found that new neurons were being generated in females as well as males, even when the females were not being given testosterone. Evidently, neuronal generation is a common feature of canaries throughout their adult lives.

New neurons were especially common in the HVc. In female canaries, new neurons were being generated at a rate of about 1.5 percent of the total number of neurons in the HVc every day. This raised an obvious problem. The loss of neurons could be accepted in males. Inventories showed that males have an average of about 41,000 neurons in their left HVc's in the spring and about 25,000 in the fall, when full song has degenerated into plastic song. But the size of the HVc remains approximately constant in females. If they were generating such a large number of new neurons every day, neuronal death must be common in their brains as well as in the males' brains.

Such a loss of neurons was unprecedented in any other animal. The HVc and other parts of the forebrain in canaries are assumed to contain memories of songs and other behaviors needed for canaries to survive. The constant turnover of neurons suggested either that canaries were constantly losing memories as new neurons replaced old ones or that the new neurons were somehow replacing only those neurons not involved in memory.

Nottebohm speculates that in fact old memories may be lost but with a corresponding benefit: the newly minted neurons may offer fresh ground for new memories. This would correlate well with the habits of canaries in the wild. When the breeding season is over, pairs of birds revert to foraging singly or in flocks and no longer associate. During this period, the need to produce or remember full song is not essential. If old neurons died and previous memories faded during this period, it might free up space for the formation of new memories, as a new breeding season approached. Similarly, for females, the loss of old memories might create new space to learn the songs of males in the spring.

This plasticity in the adult brain could offer a significant advantage to canaries and other birds. "If you needed a brain that will give good service over a ten-year period, you might need a much larger brain. But if you have replaceable modules, then you can make do with a smaller brain, which is particularly important with an airborne creature that has to lug a brain around wherever it goes," Nottebohm observes.

"This is strictly hypothetical, and as with most hypotheses in science cannot be expected to survive for more than a few years. But at this time it seems like a hypothesis with some merits to it, in that it focuses on the nucleus of the cell itself for long-term memory, which is a direction others are also going in."

The Birthplace of New Neurons

The next logical question is where the new neurons in the canaries' brains are coming from. Nottebohm and his team quickly ruled out the possibility that existing neurons were dividing. "That part of the dogma remains intact," he says. Instead, neuronal replacement in the brain of adult canaries seems to draw upon a mechanism normally active only during development.

The brains of vertebrates take shape in a particular way. New cells are created in zones surrounding the open ventricles of the embryonic brain. Some of these cells are neurons; others are various kinds of glial cells, which will provide nutrients and other support to the neurons throughout life. Among these glial cells are a particular type known as radial glial cells. These cells extend long arms, or processes, from the ventricular zone to the outer surface of the brain. Like someone shimmying up a rope, neurons then migrate along these radial glial cells to their final destinations in the developing brain.

Normally, the radial glial cells disappear before birth, converted into other kinds of glia. But Nottebohm has found that radial glial cells persist throughout life in canary brains. His team has found a monoclonal antibody, an immune system molecule that can be made in virtually unlimited quantities in the laboratory, that selectively binds to radial glial cells. It reveals a network of radial glial processes extending outward from the ventricles of adult canaries, just as in development.

Using radioactive thymidine, Nottebohm's team has been able to demonstrate the generation and migration of neurons along these radial glial cells. If a canary is injected with radioactive thymidine and killed a day later, a rich layer of labeled cells can be seen surrounding the ventricle. If the bird is killed six days later, some of the labeled cells have started migrating away from the ventricular zone. As time goes on, there are fewer labeled cells on the walls of the ventricle and more labeled migrating cells. By the fortieth day, there are very few labeled cells left in the ventricular zone.

This still does not answer the question of exactly where the new neurons come from. Are there neuronal stem cells surrounding the ventricle that can divide and produce neurons? Or do the new neurons have some other birthplace?

Nottebohm's experiments have pointed toward an unexpected source. "We believe that the most likely candidate for neurogenesis in the adult brains are the radial glial cells themselves," Nottebohm says. "Not only do they provide a path for the migration of newly born neurons, but they also provide the parent."

"We're still trying to document this further," he adds. "We're not a hundred percent sure that all neurons are generated from radial glial cells. But this seems to be the most conservative interpretation at this time."

Implications for Other Species

Neuronal replacement has not been observed in the brains of adult humans or other primates. If a part of the brain is injured, it does not seem able to repair itself internally. In some cases, other parts of the brain may be able to take over for the damaged part, as in parietal lobe syndrome. It may also be possible to transplant tissue from an external source into the brain and have that tissue take over lost functions. One kind of tissue that may prove especially effective is neural tissue from fetuses that have not come to term because of spontaneous or elective abortions. This fetal tissue is especially adaptable because of its early stage in development and readily adjusts to its new environment. However, consideration of such transplants has been beset by ethical difficulties, because some people believe that using fetal tissue for transplants could make abortion morally more acceptable.

The absence of neuronal replacement in humans does not mean that it is impossible, according to Nottebohm. "Neurons are generated in the ventricular zone in all species during development," he says, "and quite likely the genetic control in all species is quite similar. So if you could find the genes that control this behavior and learn how to turn them on and off, it would produce a new kind of brain repair. The brain could be induced to do the work itself rather than bringing foreign cells from outside to do the job, which of course will also solve some of the ethical problems that have been worrying people."

The Evolution
of Ethics

*C*anaries sing because they are genetically programmed to do so. The detailed content of their songs may vary as a result of learning, but the general characteristics of their songs and the urge to sing both arise from the neural structures generated by their genes.

Do the same forces drive human behavior? Do we behave in certain ways, including moral ways, because of genetically determined predispositions laid down over millions of years of human evolution? Or is our social behavior shaped only by learning, by the influence of our cultural environment on an initially blank and essentially malleable human mind?

Biologists and philosophers have been examining the connections between evolution and ethics since the time of Darwin. As evolutionary theories of human origins began to replace religious precepts, ethicists likewise turned to evolutionary ideas in the search for justifications of ethical systems. One of the earliest and most ardent of these evolutionary ethicists was Darwin's contemporary Herbert Spencer. For Spencer, the competitive aspects of evolution were paramount. Darwin's theory of evolution recognizes that more organisms are born than can survive and reproduce. Individuals therefore compete for resources and reproductive success. The winners in this competition pass on their genes to the next generation, which tends to perpetuate the traits that led to success. In the words of Darwin's theory, these traits have been "naturally selected."

Spencer and his philosophical successors extended this view to the social and economic realms. They advocated an extreme form of economic *laissez-faire*, in which the strong should prosper and the weak be brushed aside. In its various guises, this philosophy has come to be known as Social Darwinism, though it owes more to Spencer than to Darwin. In fact, it was Spencer, not Darwin, who coined the phrase "survival of the fittest."

Normative Ethics and Metaethics

According to Michael Ruse, professor of history and philosophy at Guelph University in Ontario, any philosophical system such as Social Darwinism must satisfy two sets of demands: those of normative ethics and those of metaethics. Normative ethics is concerned with what a person should do, with the rules and guidelines by which people should live their lives. For instance, the ten commandments are normative ethics: thou shalt not kill.

Metaethics is concerned with why a person should follow those rules, with the justification for ethical codes of conduct. In the case of the ten commandments, the justification may be that God wishes a person to follow those rules. Many other justifications for ethical systems are also possible. For instance, utilitarians might contend that one ought to maximize happiness for the greatest number. Kantians might say that one ought to treat people as ends in themselves and not as means to the ends of others.

At the normative level, Social Darwinism advocates that people acquiesce to or even encourage the workings of evolution, as expressed in its competitive nature. This normative principle can be interpreted in many ways, Ruse points out. Some people may use it to argue for free market economies in which the government plays little or no role. For instance, as prime minister of Britain, Margaret Thatcher has been reported as saying, "What we need are Darwinian principles."

Others may interpret Social Darwinism in different ways. One of Andrew Carnegie's philanthropic interests was the founding of libraries. As a Social Darwinian, Carnegie felt that libraries were a place where the poor but bright child could go to work his way up in society.

As Ruse points out, to be fully persuasive these normative principles must be buttressed by metaethical justifications. In the case of Social Darwinism, this justification is straightforward: an abiding belief in evolutionary progress. According to this view, evolution inherently generates value by leading from simple to complex, from lower to higher. Humans, in this light, are the end result of evolution, the highest point on the evolutionary scale.

Competition and Cooperation

Social Darwinism can be faulted on a number of grounds, according to Ruse. For one thing, in its most competitive forms, it is simply immoral. "Morality does not consist in walking over the weak and the sick, the very young and the very old," Ruse says. "Someone who tells you otherwise is an ethical cretin."

Biologists have also long since given up the idea of evolution as a generator of progress. Darwin himself pointed out the fallacy of calling one organism higher and another lower. To the extent that evolution acts to produce organisms well adapted to their environments, a bacterium or a virus is as successful as a human being (and perhaps more successful). It is self-centered to think of ourselves as the end product of evolution, an anthropomorphism made ironic by our brief tenure on the planet. We are the first species that has been able to describe evolution, but that does not mean that evolution has been engaged for over three billion years in the construction of human beings.

Biologists have also come to have a much broader view of evolution than that espoused by Spencer and other Social Darwinians. Especially in the last few decades, they have explored the many kinds of behavior besides competition that confer evolutionary advantage. For instance, if a fight over a piece of food leaves both combatants so weary that they are easy marks for other competitors, neither benefits. But if they share that piece of food, both may be better off.

Biologists have found that this kind of cooperative behavior is widespread in the animal world (it may even exist in the plant world, Ruse points out). Cooperative behavior in nonhuman species has come to be known as altruism, though strictly speaking it is not a moral act. Instead, it is genetically programmed into animals, a part of their evolutionary inheritance.

There are several mechanisms by which such altruism can arise. One is simple reciprocation: one animal might assist another with the expectation that sometime in the future the favor will be returned. If this behavior furthers reproductive success, it will tend to be perpetuated. Another kind of altruism is kin-related. If an animal helps a genetically related animal to succeed, part of the altruist's genes will be passed on to future generations. For instance, since brothers and sisters share half of their genes, assistance to a sibling can further the genes responsible for that altruism, even if the altruist does not benefit.

The idea that genetically determined altruism shapes human behavior accords well with our everyday experience, according to Ruse. We are surrounded by acts of human altruism, from Mother Theresa's ministrations to the care of parents for their children. There are also good reasons to believe that cooperative behavior must have been a valuable trait in prehistoric humans. Our ancestors were less agile and strong than the animals that were often their prey. High degrees of cooperation, fostered by human intelligence and tight social bonds, would often have been essential.

The Kinds of Kindness

Altruism could occur in human beings in one of several ways, according to Ruse. One possibility is that it is hardwired into our genes, as it is for ants and other animals. In this case, we would have no choice but to be altruistic; our natures would demand it.

But this mechanism is not well suited for humans. Our lives are so socially complex that flexibility is a necessity. Genetically programmed altruism—or any other kind of rigid social behavior, for that matter—would not be creative enough to deal with constantly new and changing conditions.

Another possibility is that we consciously or unconsciously calculate the evolutionary advantage to ourselves when faced with the choice of helping another person. Essentially, this posits a kind of ethical "superbrain" that would rationally guide our actions.

But this, too, is unlikely, Ruse contends. The human brain is too slow to calculate such odds when presented with a typically human situation. "By the time the calculations are made, the opportunity is long past."

The most likely possibility, according to Ruse, is that our altruistic tendencies arise from certain strategies or innate ideas that are part of our genetic heritage. In this view, we reason and feel in certain genetically determined patterns. Culture undoubtedly modifies these patterns, in some cases magnifying them (as in the case of sexual differences) or diminishing them (as with aggression). But the tendencies themselves are biological, not cultural.

This does not mean that our actions are dictated by our genes. Almost all of us violate ethical standards some of the time, and some people violate these standards most of the time. To be moral creatures, we must have the option to chose or not chose to be moral. A view of human altruism as rooted in biological altruism does not deny free will, says Ruse, but it does influence how we think about free will.

Evolutionary Justice

As a philosophical system, a morality rooted in biological altruism must provide both a normative ethics and a metaethics. In the area of normative ethics, Ruse calls upon the ideas of the moral philosopher John Rawls. In his book *A Theory of Justice*, Rawls considers the following question: What kind of society would a reasonable person design if that person did not know what position he or she would occupy in that society? Such a society is not necessarily one in which everyone is equal. If doctors were paid as much as everyone else, few people would care to invest the time and commitment needed to become good doctors. Instead of an equal society, Rawls suggests, a reasonable person would want a society that was fair. Freedom and liberty would be maximized, and society's rewards would be distributed so that everyone benefits as much as possible within the constraints of fairness.

This is exactly the kind of normative ethics that biological evolution would produce, according to Ruse. It is a system in which people who naturally tend to look after themselves behave ethically. Rawls himself points toward the correspondences between a fair society and evolutionary principles, though he goes on to try to base such a society on a different foundation. But at a normative level, there is little difference between the workings of evolution and Rawls' conception of fairness.

What about the metaethical foundations for a biological morality, Ruse asks? It is one thing to say that people ought to treat each other fairly; it is quite another to say why they should do so.

According to Ruse, this metaethical justification is not what one would expect. "I'm afraid the justification is going to disappoint the traditional evolutionary ethicist, because what one finds is that ethics has no justification." We believe that we should behave morally, Ruse asserts, but ultimately we cannot offer a reason for this belief. Philosophers may search for an objective, rational grounds for ethics; others may root their ethics in the desires of God. But there is no extrinsic basis to ethics, Ruse contends. It is part of our subjective psychology, a sort of "collective illusion," generated by natural selection to promote reproductive success.

However, the fact that we believe ethics to have an objective foundation is critical, Ruse explains. Ethics only works because everyone believes in it. If people were willing to forgo morality because they believed it to be an illusion, the system would break down. "Ethics is not objective, but it only works if we think that it is objective," Ruse says. "In other words, my case is only plausible if you find my conclusions profoundly unsatisfying."

Implications of a Biological Morality

An ethical system based on biological morality differs in some respects from traditional ethical systems, including Christianity, Ruse points out. For instance, some theologians would argue that Christ's admonition to forgive your brother seventy times seven means that Christians should continually forgive the transgressions of others, even if that behavior does not change. But evolutionary ethicists would say that there is a limit to how much someone can be forgiven, Ruse observes. A person who continually violates the ethical standards of a society can weaken the bond of reciprocal altruism, and such behavior cannot be tolerated indefinitely.

Biological morality also suggests that a person's moral obligation is reduced as one gets farther from oneself. "One has greater obligations to one's immediate family and friends than one has to strangers, and a great obligation to one's own society than one has to other societies," Ruse says. "I think a lot of Christians would oppose this, but I don't think my position is immoral."

Many people may hunger for a more objective basis to ethics, Ruse acknowledges. But the morality that evolution has spawned is a powerful force, because the subjective nature of that morality guarantees its hold upon human beings. "I don't think we establish ethics," he says, "I think ethics is thrust upon us. At a certain very fundamental level, I don't think we have any choice."

4 Evolution and the Biosphere

The topics covered in the first three chapters of this book—genetics, development, and neuroscience—can all be seen in two complementary lights. On the one hand, they are biological processes, complete and consistent in themselves. Biologists may not totally understand these processes, but they can study them and try to explain them in biological terms. This is the *how* of biology: how is a given biological system constructed, how does it function?

At the same time, biological processes are the products of evolution, with antecedents billions of years old. Biologists studying these processes from an evolutionary perspective are trying to answer the *why* of biology: why has a given system evolved as it has, what historical forces account for its nature?

It would be a mistake, however, to partition biological topics into the categories of "how it works" and "why it evolved." While the actions of molecules may be the most immediate explanation of a biological process, evolution is the ultimate explanation of that process. This is what makes evolution the primary unifying theme of biology. It accounts for the simultaneous diversity and unity of life, for the differences and similarities observed in organisms. No other modern idea has done so much to change our view of the biological world and our place in that world. To cite the title of a well-known essay by the prominent twentieth-century geneticist Theodosius Dobzhansky, "Nothing in biology makes sense except in the light of evolution."

Darwin's theory of evolution has been refined and strengthened since he first proposed it in 1859. Yet the central concept of Darwin's theory remains the cornerstone of modern evolutionary theory—natural selection. Darwin recognized that natural selection requires two contrasting forces. The first is a mechanism to generate the tremendous amount of variation that he observed among members of a species. The second is a process whereby some individuals succeed in passing on their genes to the next generation and others do not. In this way, nature selects traits more fitted to the environment, and those traits tend to be perpetuated. The steady accumulation of changing traits over long periods of time, combined with changes in the environment, is the substance of evolution.

Darwin did not know the mechanisms responsible for individual variation among members of a species. But the development of the science of genetics in the twentieth century has justified and clarified many of his assumptions. Today we know that individual variation arises from mutations and rearrangements of the genes, the sequences of nucleotides that code for proteins. This variability is so extensive that no two human beings (with the exception of identical twins) are likely to have ever had identical genomes.

With the explosion of molecular biology since the 1950s, the study of evolution has progressed to the molecular level. This advance has added a new and largely unanticipated level of complexity to evolutionary theory. Sequencing of genes and proteins has revealed much more variation at the molecular level than biologists had expected. Genes have been found to rearrange themselves and transfer between organisms in ways that were previously unknown. Molecular biology has revealed that relatively little of the genome in complex organisms codes for proteins, raising the question of what role the noncoding portions of the genome play in evolution. It has also demonstrated the evolutionary importance of the regulatory regions of DNA, since many species are distinguished not so much by the proteins they produce as by the amount of those proteins and the timing of their production.

Molecular biology has also had a large impact on more classical studies of evolution. For instance, it has made important contributions to the field of systematics—the classification of organisms and description of their relationships. Analyses of DNA or protein sequences can reveal differences between organisms too subtle to discern in outward appearances or behavior.

Molecular biology has also come to the aid of field studies. It has been used to study the structure and activities of populations, by tracing the flow of genes through interbreeding groups. The formation of new species—still a prominent concern in evolutionary biology—can now be studied on a genetic level. The evolutionary history of microorganisms, once a murky area in biology, has become much clearer because of the application of molecular techniques.

There are many aspects of evolutionary biology that cannot be explained on the molecular level. Natural selection, for example, is not a molecular process; it is rather the result of interactions among different organisms within a complex environment. In general, explanations in evolutionary biology cannot all be reduced to a single organizational level. They have to call upon many different levels.

These levels of organization in biology extend up to the broadest level of all—the entire earth and its collection of living things. Organisms do not only adapt to the environment; they also change the environment, and in doing so they change the course of evolution. Human beings are not the first creatures that have had the ability to remake the entire planet and its biosphere. It has been done before, by organisms much more humble than us. As shown later in this chapter, we would not be here if not for these organisms. And as shown in the essay that follows this chapter, human beings have not taken this interdependence of the biosphere to heart in their treatment of the earth's other species.

The Evolution of Proteins

One of the most remarkable accomplishments of molecular biology has been to trace the course of evolution in single molecules. According to Francisco Ayala, professor of ecology and evolutionary biology at the University of California at Irvine, this technique "has truly revolutionized the reconstruction of evolutionary history."

One of Darwin's boldest and most controversial conjectures was that all organisms are descended from common ancestors. In other words, if any two organisms living today could be traced back through evolutionary time, at some point their lines of descent would converge. The closer two organisms are in evolutionary terms, the more recent their ancestor. The most recent common ancestor of humans and chimpanzees, for example, seems to have been an ape-like creature, now extinct, that lived in Africa about 5 million years ago. The most recent ancestor of humans and mushrooms was probably a single-celled organism living in or near the water over a billion years ago.

Because of their common origins, all organisms share certain metabolic processes. We have proteins in our bodies that serve the same functions as proteins in mushrooms. However, these proteins are not identical. Over time, the DNA that codes for proteins undergoes random mutations, which supply the variation needed for evolution. These mutations can change the sequence of amino acids in a protein—replacing a glycine molecule, for instance, with one of the other 19 amino acids that commonly make up proteins. Therefore, as two evolutionary branches emerge from a common ancestor, the proteins in those branches also evolve. But because these changes are random, the proteins change in different ways. As time goes on, the proteins become more and more dissimilar.

Proteins cannot change unrestrictedly. If a protein is to serve a given function, certain amino acids must stay the same, and certain relationships among the amino acids must be maintained. If a mutation does alter a critical amino acid in a protein that is essential to an organism, the organism will not survive and the mutation will not be passed on. On the other hand, if a mutation alters the function of a protein in such a way that the fitness of the organism is enhanced, that mutation can be selected for and the mutation will spread.

Some biologists have proposed that relatively few mutations have such a positive effect. Among those mutations that do not cause a decrease in fitness, they argue, the great majority are simply neutral in their effects. Such a neutral mutation may change the sequence of amino acids in a protein, but it will not affect the protein's overall function. Other biologists disagree, contending that natural selection plays a much larger role in guiding protein evolution than the believers in the so-called neutrality theory claim.

A Molecular Clock

If the neutrality theory were correct, Ayala observes, it would have profound implications for the study of evolutionary relationships. It would mean that random changes are incorporated in DNA at a more or less constant rate. The interval between changes in a given protein would vary, in keeping with their random nature. But over a long enough time these intervals would average out.

In this way, changes in specific proteins would serve as a sort of evolutionary molecular clock. By measuring the differences in DNA or protein sequence between two different organisms, it would be possible to determine how much time has elapsed since those organisms diverged from a common ancestor.

A classic example of such a molecular clock, Ayala notes, is the protein cytochrome c. Consisting of 104 amino acids in vertebrates and a few more in some invertebrates, plants, and fungi, cytochrome c is a protein that evolved over a billion years ago to help organisms break down organic molecules and supply themselves with energy. In the 1960s, Walter Fitch and Emanuel Margoliash studied cytochrome *c* molecules from 20 different organisms, including the fungi *Neurospora* and *Candida*, the yeast *Saccharomyces*, insects, a fish, reptiles, birds, and mammals, including man. They determined the differences in amino acids among the different proteins and calculated the minimum number of substitutions that would be needed in the DNA of the organisms to account for those differences. They then used this information to arrange the organisms in evolutionary time, as shown in Figure 4-1.

The results astonished biologists. From a single molecule, Fitch and Margoliash had reproduced centuries of work by biologists in tracing evolutionary relationships among different organisms. "I can well remember in 1967 reading this paper in *Science* and being literally dumbfounded," says Ayala. "Here they were looking at a very small molecule, 104 amino acids, and in the midst of a single molecule the whole of evolutionary history could be reconstructed by and large correctly. Furthermore, there are tens of thousands of genes coding for proteins, and each one of these genes or each one of these proteins could provide us with a molecular clock, and therefore with an independent reconstruction of evolutionary history. This is what caused the enormous enthusiasms of evolutionary biologists and others with this hypothesis."

However, the diagram produced by Fitch and Margoliash is not perfect. It shows turtles, a reptile, being closer to birds than to the rattlesnake, another reptile. Within the birds, chickens appear to be more closely related to penguins, whereas they are in fact more closely related to ducks and pigeons. Finally, the primates, including humans, seem to have branched away from the mammalian limb before the kangaroo did, whereas in fact just the opposite occurred. Could it be, Ayala asks, that these mistakes point toward more serious problems with molecular clocks?

An Erratic Clock

One way in which Ayala and his colleagues have been studying this problem is by analyzing a protein known as superoxide dismutase. A protein involved in protecting cells from the reactive effects of oxygen, superoxide dismutase consists of two identical subunits of 153 amino acids in humans, horses, yeast, and mold and 151 amino acids in rats, cows, swordfish, and fruit flies. A total of 55 of the 153 possible amino acid sites are identical in all eight organisms, indicating that parts of the protein need to be conserved to maintain its function. The other 98 sites vary from species to species.

The amino acid differences between humans and the other three mammals—rats, cows, and horses—are fairly uniform, ranging from 25 to 30 amino acids (Figure 4-2). This makes sense, since the fossil record indicates that primates diverged from rodents about 63 million

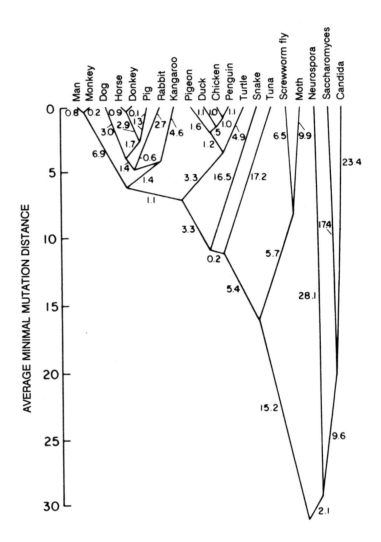

FIGURE 4-1 Differences in the amino acid sequence of the enzyme cytochrome *c* can be used to order the evolutionary relationships among organisms. The numbers between each branch point are the minimum number of changes in the nucleotide sequence of DNA needed to account for the observed amino acid differences. Reprinted, with permission, from F. J. Ayala, The Journal of Heredity 77:226–235, 1986. © 1986 American Genetic Association.

years ago, that the lineages leading to cows and horses diverged around the same time, and that the common ancestor to the four mammals lived about 75 million years ago.

The difference between humans and fish is about 48 amino acids, or nearly twice as much as between humans and the other mammals. If superoxide dismutase were an accurate molecular clock, this would mean that fish and humans diverged about twice as long ago as humans and the other mammals, or approximately 150 million years ago. However, the fossil record indicates otherwise: the most common ancestor to humans and fish seems to have lived about 450 million years ago.

The situation is even worse for the other organisms. Human superoxide dismutase differs by 69 amino acids from the superoxide dis-

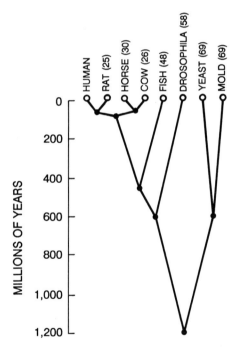

FIGURE 4-2 The evolutionary relationships derived from the fossil record for eight organisms can be compared with the differences in amino acid sequences for the superoxide dismutase enzymes they contain. For instance, only 25 of the 153 amino acids in the superoxide dismutases of humans and rats differ, whereas 69 of them are different in the respective enzymes of humans and yeast. The fungi lineage split from the line leading to human beings about 1.2 billion years ago, the insects about 600 million years ago, the fish about 450 million years ago, the ungulates about 75 million years ago, and the rodents about 63 million years ago. Reprinted, with permission, from F. J. Ayala, The Journal of Heredity 77:226–235, 1986. © 1986 American Genetic Association.

mutase of yeast and molds, or about three times as much as within the mammals. Yet the lines leading to humans and to yeast became distinct an estimated 1.2 billion years ago, 20 times more distant than the common ancestor among the mammals.

The raw numbers of amino acid differences are not a completely accurate measure of evolutionary distance, Ayala points out. For instance, if an amino acid mutated at some point during evolution and then mutated back to the original amino acid, the double substitution would be hidden. Various statistical corrections can be applied to the numbers to correct for such events. But these corrections are not nearly enough to account for the discrepancies in the substitution rate of superoxide dismutase.

Another possibility is that the amino acid substitutions in superoxide dismutase are constrained in some way that is not yet thoroughly understood. For instance, there are 98 variable sites in the superoxide dismutases of the eight organisms studied. Perhaps this value approximates a maximum number of differences and the rate of amino acid substitutions slows down as it approaches that value. However, the greatest number of substitutions between any two organisms is only 69, nowhere near the maximum. Also, a maximum number of substitutions does not seem to be a factor with comparable proteins, like cytochrome c. Overall, Ayala says, one is forced to conclude that superoxide dismutase "is not a very good molecular clock."

The Value of Molecular Clocks

"What is more common," Ayala asks, "the relative regularity of cytochrome c or the capriciousness of superoxide dismutase? I think the answer is that we don't know. The amount of information that we have is very limited. If the situation that we have with superoxide dismutase prevails, we are not going to be able to use the evolutionary clock very readily, unless we become more sophisticated about it and learn some things."

Nevertheless, there are certain circumstances in which molecular clocks can still provide valuable results, Ayala contends. For instance, some proteins, such as cytochrome c, appear to be better clocks than others. By comparing many such clocks, biologists can learn which proteins provide the most valid results. Such information will become increasingly available as DNA sequencing projects for humans and other species get under way.

Individual molecular clocks can also be valuable within certain limits. By looking at relatively similar groups of organisms, in which evolutionary processes are presumably similar, and by considering relatively long periods of time, in which variations in the clock can average out, molecular clocks can yield accurate results. Even superoxide dismutase is a good molecular clock if one looks just at the evolution of mammals. But to apply such clocks more widely, Ayala concludes, biologists need to learn more about the factors that shape protein evolution.

Two Periods of Life

Based on the geologic record, the history of the earth can be divided into two very unequal periods (Figure 4-3). About 600 million years

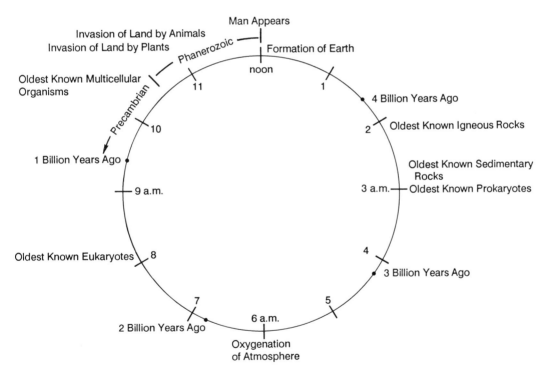

FIGURE 4-3   Organisms capable of leaving easily visible fossils evolved only in the last one eighth of the earth's history. Fossils of soft-bodied multicellular organisms have also been found, but they extend back to only about 700 million years. Before that the earth was entirely the province of single-celled organisms. Our species, *Homo sapiens*, appeared an estimated 500,000 years ago, which amounts to little more than 4 seconds on the evolutionary clock.

ago, fossils suddenly appear in great profusion in sedimentary rocks—first marine plants and animals, and later their terrestrial descendants. The period since then is known as the Phanerozoic, from the Greek word for visible or manifest.

The first geologic period in the Phanerozoic is known as the Cambrian. Everything before the Phanerozoic is therefore known simply as the Precambrian. The earth seems to have coalesced from a cloud of matter circling the sun 4.5 billion years ago. So the Precambrian period on earth spans about seven-eighths of its history.

For well over a century, biologists searched without success for conclusive evidence of fossils in Precambrian rocks. The sudden appearance of living things 600 million years ago posed a serious problem for the creators of evolutionary theory, who believed that all organisms arose through a process of gradual change from other organisms. In *On the Origin of Species*, Darwin wrote, "To the question why we do not find rich fossiliferous deposits belonging to . . . periods prior to the Cambrian system, I can give no satisfactory answer. . . . The case at present must remain inexplicable; and may be truly urged as a valid argument against the views here entertained."

Only in the past few decades has the puzzle finally been solved, according to J. William Schopf, professor of paleobiology at the University of California at Los Angeles. It is not the case that fossils do not exist in Precambrian rocks. The problem was that people were looking for the wrong kind of fossils.

The key to the puzzle of Precambrian life begins with a certain kind of rock deposit known as a stromatolite (Figure 4-4). Shaped like a stack of mattresses (or *stromas* in Greek), stromatolites were first described in the early 1800s. Almost immediately, some biologists began to speculate whether they might have been formed by living organisms. But they contain no visible fossils, and most biologists concluded that stromatolites were caused by nonbiological processes. "This debate went on and on, and many geologists simply refused to investigate these structures, largely because they had been taught by their professors that they weren't worth the time," says Schopf. "But they turn out to be worth the time."

Several things came together to change biologists' minds about the origins of stromatolites. For one thing, biologists began to find and examine colonies of bacteria that produce structures remarkably like the stromatolites seen in the geologic record. In a few dry, salty, and sunny places in the world, such as the coasts of Baja California and northwestern Australia, bacteria grow in columns rising from shallow water (Figure 4-5). Typically these columns are not hard; they can be

FIGURE 4-4 Stromatolites in the fossil record typically resemble stacked pancakes rising in pillars or mounds. These specimens occur in limestone deposits about 1.3 billion years old from Glacier National Park, Montana. Photograph courtesy of J. William Schopf.

FIGURE 4-5 One of the few places in the world where stromatolites still live is in Shark Bay, Western Australia. For much of its history, the earth's surface probably looked something like this. Photograph courtesy of J. William Schopf.

cut with a machete, Schopf says. But in some places, conditions are such that calcium carbonate, the mineral constituent of limestone, adheres to their surfaces, forming a hard pillow-shaped structure. If covered with sediments and compacted, these deposits would be virtually indistinguishable from geologic stromatolites.

Living stromatolites are complex ecosystems composed of several different kinds of bacteria. The top layer consists of a type of bacteria known as cyanobacteria, the same kind of bacteria responsible for pond scum on bodies of stagnant water. Like all bacteria, the cyanobacteria are prokaryotes—that is, they do not have a distinct nucleus containing their DNA. For this reason, a more common name for cyanobacteria—blue-green algae—is not entirely accurate, since all true algae have nuclei and are therefore eukaryotes. The cyanobacteria are also photosynthesizers, like green plants. They use light from the sun to convert carbon dioxide and water into the organic compounds that they use to grow.

These compounds in turn support different kinds of bacteria that live beneath the cyanobacteria in modern stromatolites. These bacteria get their energy as animals do, by feeding off the organic molecules produced by photosynthesizers.

Schopf points out that modern stromatolites might seem to lead a rather precarious existence. "They grow near the sediment-water interface," he says, "which is not a good place for organisms that rely on the sun to exist. They can get buried when the spring rains come and detritus covers them up. They can no longer see the sun and should die."

Modern stromatolites get around this problem by growing. The cyanobacteria making up the top layer of the column are phototactic—they move in response to light. "So indeed they do get buried, but as they do they glide up through the accumulated detritus and make a new layer," Schopf explains. "And there they sit, happy as clams, or happy as blue-green algae, until they get buried again and again, layer after layer. And that's how a stromatolite forms."

The discovery of modern stromatolites was a suggestive piece of evidence, but it was not enough to prove that the stromatolites in the geologic record were also made by bacteria. To do that, biologists had to find evidence of the organisms that built the geologic stromatolites. They did this by examining stromatolites preserved in silicon dioxide—the mineral quartz. If such stromatolites are sliced with a diamond-impregnated saw and ground down by hand, it is possible to make thin layers that can be seen through with a microscope. By exhaustively searching through such translucent layers, researchers managed to find the remains of fossilized cells (Figure 4-6). Sometimes the cells are joined into rows, or flattened into disks, just like the bacteria in modern stromatolites. The fossils of the Precambrian era had finally been found.

It was immediately obvious why earlier biologists had found no evidence of Precambrian life. "The Precambrian was the age of microscopic life," Schopf points out. "People were asking the wrong question. They were looking for macroscopic organisms, for the equivalent of trilobites and clams. But such organisms had not yet evolved. The planet was dominated for three billion years by microbes, which set the stage for all subsequent evolution."

Bacteria and the Atmosphere

Stromatolites first appear in the geologic record about 3.5 billion years ago, and already they contain fossilized cells (this is the earliest

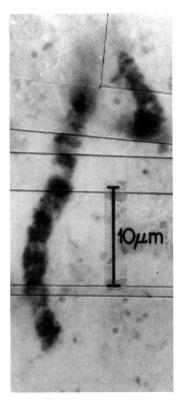

FIGURE 4-6   Fossilized cells from 3.5-billion-year-old deposits in northwestern Australia still exhibit cell walls and a filamentous form. These fossils are among the oldest now known. Photomicrograph courtesy of J. William Schopf.

direct evidence for life on earth, as described in the box below). But what kinds of cells were they? Were they photosynthesizing bacteria, like the ones that form stromatolites today? Or did some other kind of organism form these early stromatolites?

To search for evidence of photosynthesis, paleobiologists have studied the rocks in which these early cells are embedded. Carbon dioxide, which photosynthesizers absorb to create organic compounds, has always been abundant in the earth's atmosphere, because it spews into the atmosphere and oceans from volcanoes and deep-sea vents. But not all carbon dioxide is alike. The carbon in carbon dioxide consists of two different isotopes—carbon-12 (with six protons and six neutrons) and carbon-13 (with six protons and seven neutrons). When organisms photosynthesize, they tend to slightly prefer carbon-12 to carbon-13.

....

Extinction and the Fossil Record

Molecular clocks can only be used with living creatures and those few organisms for which perishable DNA or protein samples have been preserved. For creatures that are now extinct, biologists must rely on the fossil record to derive evolutionary relationships.

Fewer than one percent of all the species that have ever lived on the earth still exist today. As described in the essay following this chapter, there are somewhere between 5 and 30 million species now living on the earth, of which somewhat more than a million have been named and described. It is difficult to estimate the number of species that have lived on the earth throughout its 4.5-billion-year history, but it is thought to be somewhere around 4 billion. For reasons that are not yet clear, major extinction events seem to occur about every 26 to 28 million years. Some biologists believe that periodic changes in climate or sea level are responsible for these events, whereas others blame them on collisions with extraterrestrial objects.

Paleobiologists have described some 250,000 extinct species of plants, animals,

and microorganisms occurring in the 3.5 billion years during which living things are known to have existed. Museums and universities have collections containing tens of millions of fossilized specimens, but relatively few have been thoroughly examined.

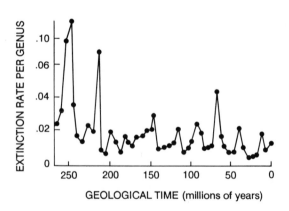

Mass extinctions seem to occur about every 26 million years, although the reasons for this periodicity are still obscure. Reprinted, with permission, from D. Kerr, "If There Was One Killer Impact, Were There More?" Science 242:1381, 1988. © 1988 American Association for the Advancement of Science.

Therefore, when these organisms die and are fossilized, the carbon in those fossils tends to be enriched in carbon-12 compared with the carbon in nonbiological deposits.

Researchers have measured the ratio of carbon-12 to carbon-13 in early stromatolites, and the results indicate a definite enrichment in carbon-12. "The carbon isotopes are consistent with the presence of photosynthesis going back three and a half billion years ago," says Schopf.

The question then becomes what kind of photosynthesis the early bacteria engaged in. Plants and cyanobacteria combine carbon dioxide with water to get organic molecules, giving off oxygen in the process. But other kinds of photosynthesis are possible. In particular, certain bacteria that evolved before cyanobacteria use hydrogen sulfide rather than water in photosynthesis, releasing sulfur as a waste product. The bond between sulfur and hydrogen is easier to break than the one between oxygen and hydrogen, and the biochemical machinery needed is less involved. Perhaps these bacteria were responsible for the early stromatolites.

Schopf and his colleagues have attacked this question by examining the size and shape of the fossils found in the earliest known stromatolites. They are larger than modern sulfur-producing bacteria, he points out, being more along the lines of modern cyanobacteria. They are also organized into rows or globular colonies encased by thick, layered sheaths, a common feature among cyanobacteria and rare among other prokaryotes. Altogether, he says, the evidence supports the idea that oxygen-producing bacteria had evolved by 3.5 billion years ago, but it is not conclusive.

To examine the issue of when oxygen-producing photosynthesizers first appear, paleobiologists again turn to the geologic record. When photosynthesizers give off oxygen, it enters the atmosphere and oceans. There it influences the formation of certain minerals, leaving a trace of its presence. By examining these minerals, it is possible to get a fairly good idea of when oxygen first appeared in abundance in the earth's atmosphere.

The best evidence for atmospheric oxygen comes from a geologic feature known as banded-iron formations. Between about 3 billion and 2 billion years ago, a series of red bands formed by the mineral hematite, or iron oxide, appear in the geologic record. "No matter where you go on the earth's surface, 2 to 3 billion years ago, you're going to find this type of rock," says Schopf. These bands appear to have been formed by deposition in the earth's oceans. Before the earth's atmosphere contained oxygen, the oceans were saturated with ferrous iron, which can exist dissolved in water. But as photosynthetic bacteria began giving off oxygen in large quantities, this oxygen combined with ferrous iron to produce ferric oxides, which are insoluble in water and dropped to the ocean floor. There they were compacted by other sediments to form the banded-iron formations.

Today, most of the world's commercially important deposits of iron come from these formations. "Why were there steel mills in Pittsburgh, or an automobile industry in Michigan?" Schopf asks. "Because that's where the iron was processed." This iron was deposited as banded iron formations 2.2 billion years ago, along the shores of an ancient sea. As Schopf says, "The world rusted."

Stromatolites appear in abundance in the geologic record about 2.8 billion years ago. It is reasonable to assume, Schopf asserts, that they were built by the same oxygen-producing photosynthesizers responsible for the banded-iron formations.

For several hundred million years, iron in the oceans and other geologic sinks absorbed the oxygen given off by these early photosynthesizers. But eventually these sinks for oxygen got used up. At that point, oxygen began to accumulate in the atmosphere. The result was a biological revolution that would forever change the course of evolution.

An Oxygenic Atmosphere

The introduction of oxygen into the atmosphere was one of the most momentous events in earth's history. Today the earth's atmosphere is about a fifth oxygen, with virtually all of that oxygen produced biologically by green plants and cyanobacteria. But this abundance of oxygen disguises the fact that it can be a deadly toxin to life. Oxygen reacts with organic molecules, destroying the functions of proteins, nucleic acids, and other essential molecules. Essentially, oxygen burns these substances, removing their biological activity.

The development of an oxygenic atmosphere undoubtedly drove many organisms to extinction. Others found ways to avoid oxygen by retreating to anaerobic, or oxygenless, environments. Still others managed to develop biochemical defenses against the reactivity of oxygen. For instance, superoxide dismutase, the subject of Ayala's molecular clock studies, evolved to protect intracellular systems from toxic derivatives of oxygen.

But evolution is endlessly opportunistic, and while some organisms suffered because of oxygen, others thrived. In the upper atmosphere, oxygen atoms formed a layer of ozone, which absorbs ultraviolet light and keeps it from reaching the earth's surface. Previously, organisms had been forced to protect themselves from the biologically damaging effects of this radiation by shielding themselves from direct sunlight or evolving elaborate biochemical defenses. With the ultraviolet light all but gone, they could spread into ecological niches that had been closed to them. (The shutting off of ultraviolet light also eliminated the plentiful source of energy that many biologists believe contributed to the formation of life, as described in the box below.)

Even more important, organisms began to develop ways to use oxygen to their advantage. They evolved biochemical mechanisms that used oxygen to break down foodstuffs, resulting in a much more efficient use of organic molecules to get energy. They developed biochemical pathways in which oxygen was an essential participant, resulting in such molecules as steroids, carotenoids, and unsaturated fatty acids. These organisms could not only tolerate the presence of atmosphere; they used its growing abundance to establish their biological dominance.

The development of an oxygenic atmosphere by about 1.7 billion years ago set in motion an entirely new era in evolution. By as early

....

The Origins of Life

The fossil record contains many gaps, but none is longer or more exasperating than the very first one. The oldest known rocks date from about 3.8 billion years ago, when the earth was just 700 million years old. But the first sedimentary rocks, in which fossils might be found, are about 3.5 billion years old. Already these rocks contain fossilized bacteria, so earlier forms of life must have arisen sometime in the earth's first billion years.

To learn more about evolution during earth's early history, biologists have tried to understand the chemical conditions that then existed. The earth's environment was much different then than it is today. Almost all of the planet was probably covered by water, and the atmosphere consisted of gases recently expelled by the earth's cooling rocks. Most important, the atmosphere contained little or no oxygen. Without oxygen there was no ozone layer to block the sun's ultraviolet radiation, and it fell freely upon the earth. This energetic radiation could have provided the energy to synthesize many organic compounds from molecules like water, carbon dioxide, ammonia, and methane in the earth's early atmosphere and oceans. These organic compounds, in turn, could have associated inside droplets encased within a membrane-like skin and engaged in a rudimentary biology.

It is a long way from associating organic compounds to self-replicating nucleic acids and proteins, and biologists have few clues as to how this transformation occurred. But a broad account of how life developed from inanimate compounds on the primitive earth is not unattainable. Reconstructions of the earth's early environment, biochemical studies of existing organisms, and even observations of other planets in the solar system could all shed light on life's origins.

as 1.5 billion years ago, cells with nuclei and other internal structures began to appear. To this day, almost all of these eukaryotic cells are aerobic, requiring the presence of oxygen, and the exceptions are clearly descended from earlier aerobic eukaryotes. About 700 million years ago, these cells began to form integrated multicellular colonies, and individual cells acquired specialized functions. At first these multicellular organisms had soft bodies that were rarely preserved as fossils. But about 600 million years ago organisms began to evolve hard skeletons and other body parts, which when buried by sediments left easily visible traces. Today, all multicellular organisms, including all plants and animals, are composed of eukaryotic cells.

A Modern View

The anaerobic portion of earth's history—from its creation 4.5 billion years ago until about 2 billion years ago—may seem a rather sedate time in the history of evolution. But that is because evolution tends to be equated with changes in the shape of organisms. During the first half of earth's history, the dramatic events in evolution occurred inside cells. The period was characterized by what Schopf calls a "Volkswagen syndrome"—the tendency for outward appearances to remain the same while internal mechanisms are undergoing substantial change. The ancient prokaryotes developed all of the basic biochemical machinery on which later life depends. And in doing so they transformed the earth's environment from one hostile to life to one in which advanced organisms could prosper.

Unraveling the threads of Precambrian evolution has been a difficult process and is far from over, says Schopf. It is an interdisciplinary undertaking, drawing on fields from throughout biology and beyond. "To address the problems of three and a half billion year old pond scum, one has to worry about such things as geology and minerology, microbiology and paleontology, organic chemistry, biochemistry, atmospheric evolution, a little bit of comparative climatology, and the history of science." Progress in the field has sometimes been slowed, Schopf says, by the tendency of scientists to focus on particular fields and ignore their interdisciplinary connections. But "nature is not compartmentalized," Schopf observes. "There's a great imperfection in our science, and I think it's a function of the way we educate our students. There's something to be said for an interdisciplinary education, and perhaps my remarks in some small way illustrate that point."

Human societies are now reaching a point where the interconnectedness of the biological and the nonbiological can no longer be ignored. By burning fossil fuels and destroying vegetation, we are increasing the amount of carbon dioxide in the earth's atmosphere. Carbon dioxide acts as a greenhouse gas, trapping infrared radiation and raising the temperature of the globe. If computer models of atmospheric processes are accurate, global temperatures will rise several degrees over the next century, shifting agricultural and ecological zones and possibly raising sea levels.

Humans are also modifying one of the influences that made the modern ecosystem possible. Industrial chemicals known as chlorofluorocarbons have been breaking down ozone molecules in the upper atmosphere, allowing more ultraviolet radiation to reach the ground. Besides increasing skin cancer rates, this increased ultraviolet radiation could eventually harm terrestrial and marine plants and animals, with untold ecological consequences.

By studying the geologic, atmospheric, oceanographic, and biologic processes that have shaped our modern world, biologists hope to learn more about how these forces will continue to interact. "The past determines the present," says Schopf, "and in the same sense the present determines the future."

Preserving Biological Diversity

*G*lobal change—the worldwide modification of the environment as a result of human activities—has become front-page news. Three broad trends seem responsible for this new-found concern with the environment. Industrial pollution, long a problem at the local level, has become national and international in scope, particularly through its contributions to acid rain. Increasing levels of greenhouse gases in the atmosphere are coinciding with a gradual warming of global temperatures, raising fears about the effects of future warmings on agriculture, rainfall, and sea levels. And an observed thinning of the ozone layer, including its virtual disappearance over Antarctica in the spring, has sparked concern that continued releases of chlorofluorocarbons could dramatically increase the ultraviolet radiation reaching the earth's surface.

However, discussions of global change often overlook another critical trend, according to Edward O. Wilson, professor of science at Harvard University. "There is a fourth horseman in the environmental apocalypse, which needs to be much more closely monitored and acted upon. Unlike the others, it is truly irreversible and hence unpredictable in its consequences. I'm speaking of the extinction of species caused by habitat destruction, especially the destruction of tropical forests."

Human beings are now causing a mass extinction that rivals any of the extinction events that have occurred in the earth's 4.5-billion-year history. Over the course of a hundred years—little more than a human lifetime—as many as half of the species living on the earth could become extinct. The biological diversity of the world is being irrevocably reduced, not through any conscious decision to reduce diversity but because no decision has been made to preserve it. "This is the folly our descendents are least likely to forgive us," Wilson believes.

Ethical beliefs inevitably shape a person's attitudes toward the extinction of species. Some people may hold with the Book of Genesis that God gave humans dominion over all living things to use as we see fit. Others may believe that we are charged with the stewardship of other species and are responsible for their welfare. Some people may see humans as a self-contained species, with no intrinsic responsibility toward other species except as they influence human welfare. Others might feel that because humans are a product of evolution, it diminishes humanity to let the other products of evolution be destroyed.

"The field biologist is impatient with these niceties of moral reasoning," says Wilson. "He is like a molecular biologist watching the laboratory burn down." Species are disappearing much too fast to withhold action until an ethical consensus emerges, Wilson says. Given the many known benefits of biological diversity and the unknown consequences of reducing that diversity, simple prudence would dictate that we act to preserve the world's biological heritage.

Measures of Diversity

"It is a remarkable fact that no one knows the amount of biological diversity in the world even to the nearest order of magnitude," Wilson points out. Biologists have named and described over 1.4 million species of all types since formal systems of classification were inaugurated in the 1750s. But except for a few well-studied categories such as flowering plants and vertebrates, many more species exist than have been named and described. Wilson estimates that there may be anywhere between 4 and 30 million species on the earth, over half of them insects.

Each of these species is an irreplaceable repository of genetic information. The estimated number of genes in various organisms are about 1,000 in bacteria, approximately 10,000 in some fungi, from 50,000 to 100,000 in humans and many other animals, and around 400,000 in many flowering plants. Moreover, the individual members of a species contain different genes and different versions of the same gene, resulting in diversity within as well as between species.

"A species is not like a molecule in a cloud of molecules," says Wilson. "It is a unique population of organisms, the terminus of a lineage that split off from the most closely related species thousands or even millions of years ago. It has been hammered and shaped into its present form by mutations and natural selection, during which certain genetic combinations survived and reproduced differentially out of an almost inconceivably large possible total."

Relatively few genes have been studied in great detail, and except for a handful of laboratory organisms the nucleotide sequences for any given organism, including humans, are largely unknown. Hence, when a species becomes extinct, the genetic information it contained is lost forever.

Diversity in the Tropics

By far the richest known collections of species in the world occur in tropical rain forests. Though such forests cover only about one-fourteenth of the world's land surface, they contain over half of the world's species. More accurately known as closed moist tropical forests, these forests typically contain three or more canopies of vegetation. The top canopy, formed by evergreen broadleaf trees, is very thick, so that little direct light reaches the forest floor. This absence of direct light reduces the amount of undergrowth, so that humans can walk through such forests with relative ease.

The diversity of living things within these forests is legendary among biologists. "Every tropical biologist has a favorite example to offer," Wilson says. "From a single leguminous tree in Peru, I recovered 43 species of ants belonging to 26 genera. That's approximately the same as the entire ant population of the British Isles or Canada." In ten plots totaling 25 acres in Borneo, one tropical biologist identified about 700 species of trees, more than the number of native tree species occurring in all of North America. A square kilometer of forest in Central or South America may contain several hundred species of birds and many thousands of species of butterflies, beetles, and other insects.

This incredible biodiversity is colliding head-on with a harsh reality of modern history: these areas are under some of the most intense development pressures of any ecosystems in the world. Most tropical forests occur in developing countries with rapidly growing populations. Already, 40 percent of the land that once supported tropical forests no longer does so because of human activities. And as population and economic pressures continue to grow, so will the pressures on the remaining tropical forests.

By the most conservative estimates, about 1 percent of the existing tropical forest is being cleared or permanently disrupted each year—an area about the size of West Virginia. Other estimates are much higher, though in the politically charged atmosphere surrounding deforestation such numbers are inevitably controversial. "The important point is that the rates are very very high, however you look at them," says Wilson.

Most of these areas are being permanently cleared to make way for agriculture. But one of the tragedies of tropical deforestation is that these lands are not particularly well suited to agriculture. The existence of lush tropical forests can give the impression of an abundant fertility. "But the existing tropical forests are not the rich fertile environments easily regenerated that most people imagine," says Wilson. "They're quite the contrary. They are what you could call wet deserts."

Most tropical forests exist on what are known as tropical red and yellow earths, which are acidic and poor in nutrients. When the trees are cut down and burnt, they release their nutrients into the soil, and for two or three years these nutrients can support crops. But after that the nutrients are used up or washed away, and agricultural yields decline precipitously without extensive use of fertilizers.

Once the forests are chopped down, it will take centuries for comparable ecosystems (minus the exterminated species) to regenerate fully. In some cases where damage is severe and the soil is particularly poor, the forests may never regenerate naturally. Tropical forests are therefore not necessarily a renewable resource, like forests in temperate areas. In many respects they are a non-renewable resource, like oil or minerals.

If current rates of deforestation continue, the tropical rain forests will be virtually gone by the beginning of the twenty-second century. However, some areas are disappearing much faster than the average and will be gone within a decade or two. The rate of deforestation is also increasing, leading many tropical biologists to place the disappearance of the rain forests well within the twenty-first century.

The amount of extinction that this destruction of habitat will cause depends on the number of species living in these ecosystems now (a number that is not yet known with certainty) and on how much of the forests can be preserved. Studies of island biogeography indicate that, as a general rule, when the area of a particular habitat is reduced by 90 percent, the number of species living in that habitat drops by half. In the tropics, however, this general rule may underestimate the true loss of species. Many tropical species occur in small geographical areas, so a relatively small loss of habitat can spell their extinction. Such habitat destruction can also reduce the genetic diversity within a species, leaving it more vulnerable to future disruptions.

Without conservation efforts on a massive scale, existing tropical forests will eventually be reduced to much less than 10 percent of their current area. It is therefore likely that more than half of the species now living in these areas will be lost.

Linking Development and Conservation

In the industrial world, development and conservation are often seen as competitors in a zero-sum game: when development wins, conservation loses. But that equation does not necessarily hold in the developing world. There, the biological wealth contained in natural environments can be a valuable source of increased human prosperity.

"Wild species in the rain forests and other natural habitats are among the most important human resources," says Wilson, "and so far the least utilized." Food production is the prime example. At present, people rely on only 15 to

20 species of plants for the great majority of their food supplies, and just three species—wheat, maize (corn), and rice—supply more than half. Yet there are at least 75,000 plants that are edible, Wilson observes, and many of them have qualities superior to those of the crops now in use.

Even if a wild plant is not grown as a crop, its qualities can be introduced into interbreeding crops through traditional breeding programs. Furthermore, using genetic engineering it should soon be possible to transfer valuable traits, such as disease or pest resistance, between plants that do not naturally interbreed.

Wild species are also a vast and largely untapped reservoir of new pharmaceuticals, fibers, petroleum substitutes, and other products. For instance, one in ten plants contains anticancer compounds of some degree of effectiveness. The rosy periwinkle of Madagascar provides two alkaloids, notes Wilson, vinblastine and vincristine, that can largely cure Hodgkin's disease and acute childhood lymphocytic leukemia. Now the basis of a $100 million a year industry, the rosy periwinkle is one of six related species on Madagascar. "The other five have largely been unstudied," Wilson notes, "and one is at the moment on the verge of extinction due to the destruction of natural habitats."

Plants are not the only wild species of potential value. Insects can act as crop pollinators, control agents for weeds, and parasites and predators of other insect pests. Bacteria, yeast, and other microorganisms can yield new medicinals, foods, and procedures of soil restoration. Proposals for how wild species can promote human welfare "fill volumes," says Wilson.

Approaches to Conservation

"You can't stop a Mexican peasant from shooting the last imperial woodpecker to feed his family, which in fact happened 15 years ago," says Wilson. "But in less desperate cases you can persuade people and governments, at least to some extent, that it is to their short-term and long-term benefit to preserve biodiversity. In the short term, they can get longer and richer yields from existing resources. In the long term, they're saving one of their national treasures."

A number of methods have been developed that can simultaneously further economic development and preserve biological diversity, Wilson notes. New methods of strip lumbering can yield income from tropical forests while preserving forest tracts. Proper agricultural management can conserve the nutrients in tropical soils, so that farmers do not have to keep moving to be able to work fertile ground. Land that has already been cleared needs to be enriched or restored to take the pressure off undeveloped land. And crops especially suited to the tropics, such as fast-growing trees that can be mowed to yield fiber and wood pulp, should find much wider use.

Governments and international development organizations need to make biodiversity a major consideration when planning and supporting development projects, Wilson observes. Encouraging steps in this direction have been taken; for instance, the U.S. Congress has mandated that programs funded by the Agency for International Development include an assessment of environmental impact. But much more needs to be done.

More innovative measures have also been proposed. Some people have advocated that the international debts of developing countries be partially forgiven if they undertake conservation projects. A similar approach is to buy the debt of developing countries at a substantial discount and use that credit to purchase land for preservation. "There are a lot of techniques that have been developed," says Wilson, "and it is not going to take an enormous amount of money in terms of foreign aid compared with what we have been contributing, for example, in military aid to many of these countries."

The Role of Biologists

Biological research will be an important complement to policy measures in preserving biological diversity. First of all, research in systematics and ecology is needed to get a better idea of the dimensions of biodiversity and the magnitude of the threat facing it. "There's clearly a need for a strong new effort in systematic biogeography to find out where the species are located, which areas are in need of protection, and where the species exist that might be put to immediate use in the economic sphere," says Wilson. "We're going to have to rebuild our museums and other institutes devoted to biodiversity studies to concentrate a lot more fieldwork out there in the real world."

Wilson is in favor of a biotic survey of every species—plant, animal, and microorganism—that exists on the earth, "a project comparable to mapping the human genome." Such a survey could help answer a number of vital questions in evolutionary biology. For instance, what accounts for the number of species on the earth? Is it due to something about the nature of the planet or to something about evolution? Why do hot spots of biological diversity exist? Can the diversity of natural systems be increased through human intervention?

The restoration of damaged ecosystems is another area in which biologists can make a major contribution. What are the best methods to promote the regeneration of a natural ecosystem? How can ecosystems be maintained in such a way as to promote diversity? These areas of biological research need to undergo substantial growth in the near future, Wilson contends.

Biology is not the only science that needs to become more involved in preserving biodiversity. Economics has traditionally had difficulty assigning value to biological diversity and other environmental assets because these entities exist outside the narrowly defined market economy. Psychology and sociology have never made serious efforts to study the relation of mental and social health to the vitality of the natural environment. In general, Wilson asserts, the social sciences need to become much more integrated into the realities of the natural environment and the uses of biodiversity.

Studies of biodiversity are unusual in science, because there is a strict time limit on when they can be done. Biologists and other scientists are in a race with time, and the competition is running ever faster as population pressures increase. "The study of biodiversity has unexpectedly gained a new urgency," Wilson says. "It has become as important to humanity now as medicine or molecular biology."

Afterword

The biological problems considered in this book have ranged from the molecular to the global, from the very specific to the very broad. Ethical problems can be arrayed on a similar scale. Some ethical issues involve the individual and the family: How should human tissues be used for transplantation? What are a person's obligations to family and friends? Other ethical issues involve the entire species: How did ethics evolve in human beings? What are the obligations of human beings to the other living things on the earth?

On this broadest scale, ethical and biological issues intertwine around a central question: Will the human race be able to survive and prosper in the years ahead? Science and technology have given us the power to greatly reduce our chances of survival, through nuclear war or the degradation of the environment. Science and technology have also given us the capacity to raise standards of living throughout the world, through better sources of food and energy, improved disease prevention and health care, and economic development. Thus the problems and possibilities facing humanity involve both scientific and ethical issues; they relate both to what we can do and to what we need to do.

"In an ethical sense, we have no choice other than to secure our own future," says Peter Raven, professor of Botany at Washington University and director of the Missouri Botanical Garden. "The central obligation that we as human beings have is to preserve the sustainable capacity of the ecosystem into which we have evolved. If we don't do that, I think we're clearly carrying out an immoral act."

Population and Wealth

The "major factor" in the pressure being placed on the global eco-system, according to Raven, is the accelerating increase of the human population. In 1930, the global population was about 2 billion people. Today it is more than 5 billion, and by the turn of the century it will be well over 6 billion. Given current trends, demographers do not expect population growth to cease until sometime in the middle of the twenty-first century, when it is estimated that the global population will be somewhere between 9 and 11 billion people. Even reaching a stable population at those levels will require sustained global and national efforts over the next 50 or 60 years.

Population increases have begun to strain many of the biological and geological systems that have long sustained human life. Only a few countries, and none of them in the tropics, are net exporters of food. As a result, about 500 million people receive less than 80 percent of the minimum diet recommended by the United Nations. Population growth also exerts pressure on what may be already limited sources of energy, water, and minerals. In many less developed countries, for instance, firewood is the fuel most often used to cook food, consuming a large portion of many families' budgets. But the great demand for firewood has caused whole forests to be denuded, and in many parts of the world people have to travel for miles to secure any firewood at all.

"There's no one who really has a clue as to whether we may not have already exceeded the carrying capacity of the earth," Raven says. "We're conducting a major global experiment, without controls, and with the implicit assumption that everything will somehow work out."

Population growth is not the only source of stress on the world's social, political, and biological systems. Another troubling problem arises from the inequitable distribution of wealth among the peoples of the world. About 25 percent of the world's people live in industrialized countries, which include the United States, Canada, Europe, the Soviet Union, Japan, Australia, New Zealand, and a few others. Yet this quarter of the world's population controls 83 percent of the world's economic product. By various measures of consumption, the industrialized world consumes 80 to 95 percent of the world's material goods. Therefore, three quarters of the world's people survive on less than a fifth of the goods produced in the world.

These statistics can be put in more human terms. Of the 2.8 billion people living in less developed countries, excluding China, 1 billion live in conditions that the World Bank defines as absolute poverty, meaning that they cannot count on finding adequate food and shelter from one day to the next. About half of these people are malnourished. Some 14 million children under the age of four, almost all of whom live in the developing world, starve to death or die of diseases related to starvation every year. The deaths of these children amount to one out of every three deaths that occur throughout the world.

The social and economic problems exacerbated by population growth and the unequal distribution of wealth are becoming increasingly familiar, Raven points out: unstable governments, inflation, war, massive international debts, and large numbers of displaced people and immigrants moving throughout the world. These problems cannot be confined to the countries where they occur. The anguish of less developed countries will be transferred to the industrialized world through a variety of social and economic links. Thus the people of the industrialized countries have reasons of self-interest as well as humanitarian reasons for confronting problems of declining per capita resources and inequitable living standards. "As long as we keep the distribution of wealth around the world stretched as taut as it is and fail to recognize our self-interest in joining other people of the world in building their standards of living to an adequate level, we are not taking a responsible attitude toward creating a sustainable ecosystem," Raven says.

Solutions to problems facing the less developed world need to be framed "in a context of social equality, of justice, of fairness," Raven maintains. We need to look for ways to extend the ethics that govern our relationships with members of our own society to other societies and the world as a whole.

Biology and the Future

Biological research will be central to the effort to create a productive and sustainable global ecosystem. It can produce better crops and domesticated animals through traditional breeding programs, through the selection of wild crops for cultivation, and soon through genetic engineering. It can replace dwindling supplies of energy and minerals with new and alternative resources. It can demonstrate the interconnections within the global biosphere, providing the knowledge necessary to manage global systems.

"The need to manage the environment collectively, to get away from our focus on the short term, is a need whose realization can be very much informed by biological knowledge," Raven says. "We must dedicate ourselves to the preservation of conditions that will allow people to exercise their moral judgments, to make their scientific calculations, to live the sorts of life that we live and have a reasonable expectation of passing lives of that sort on to our children and grandchildren."

Scientists and engineers in the developed countries have an irreplaceable role to play in using biological knowledge to meet human needs. Only 6 percent of the world's scientists and technologists live in developing countries. International collaboration is therefore critical if developing countries are to generate the expertise to manage their own resources.

As befits the wealth of an industrialized nation, the United States is now conducting more biological research than any nation ever has. This science is not just a "gimmick," Raven insists, done to satisfy a basic human urge for increased knowledge. "The science that we're doing, the technology that we're developing, the education that we're undertaking are of key importance for human survival," he says. "Better scientific understanding underlies collective human progress. . . . It is a way of improving the human condition throughout the world."

Will biological research provide the knowledge needed to build a prosperous and stable future for all human beings? That has to remain an open question. But the potential of biological research to do so is unquestioned. One of biology's many enticements is its open-endedness. It will never be complete, so long as there is life to study and biologists to study it. Biology is not just the science of what we are and of how we came to be—it is also the science of what we can become.

Index

DNA
    base pairs in human cells, 6
    control of biological development, 34, 82
    discovery of, 2
    fingerprinting, in forensic medicine, 17–18
    fluorescent labeling of, 20–21
    in fruit fly, 35
    genetic bases in, 5, 6–7, 15
    hybridization, 14, 24
    interaction with chemical environment, 38
    libraries, 15, 22–23
    manipulation of, 2; see also Recombinant DNA
    mosaics, 15
    probes, 10, 14, 15, 17–18
    public knowledge of, 29
    radiolabeling of, 20
    restriction fragment length polymorphisms in,
        15–16
    stability in culture libraries, 22
    structure of, 5, 6–7, 8
    synthesizers, 24
    universality of messages encoded in, 3
    vectors, 22, 24–27
DNA sequencing
    applications throughout biology, 21–22
    base pairs currently sequenced, 20–21
    clone requirements for, 22–23
    costs of, 22
    data base management, 20–21, 22
    funding for, 22
    methods, 20–21, 22–23
    objections to, 21–22
    public education on, 31
    rationale for, 29
    recombinant DNA technology and, 2, 15
    scope of project, 20, 22
    technology for, 22–23
Dobzhansky, Theodore, 81
Down syndrome, 8
Duchenne muscular dystrophy, 8, 14

E

Ecosystems
    economic and population pressures on, 102
    human destruction of, 98–99
    interactions in, 3, 92, 99
    population and economic pressures on, 108–109
    preservation of, 107–108
    regeneration of, 103
    stromatolites, 92
    see also Tropical forests
Electron microscopy, 4
Electrophoresis
    gel, 15, 20

    pulsed electric fields, 23
    two-dimensional, 24, 25
Embryonic development
    African clawed toad (Xenopus laevis), 39–42
    brain development during, 73
    in cell cultures, 39, 41, 45
    constraints on research, 35, 47, 49
    fate map of, 39
    fruit fly, 36–37, 42
    gastrulation, 40
    human, 3, 33–34, 42–43
    mouse, 43, 45
Enzymes
    defects in, 10
    ligases, 15
    restriction, 14–16
    role of, 7
Ethics and ethical considerations
    based on biological morality, 79–80
    breadth of, 107
    DNA fingerprinting, 17–18
    evolution and, 75–80
    extinction of species and, 100–101
    in fetal tissue transplants, 74
    in gene therapy, 7
    in genetic testing, 6–7, 17, 18–19
    in organ transplants, 47, 51
    normative ethics and metaethics, 76, 78–79
    pace of biological research and, 4
    predictions of diseases without cures, 18
    research with human embryos, 35
    social dimensions of, 49–50
    in use of human materials, 47–51
Eukaryotes, 89, 92, 98
Evolution
    of bacteria, 93–96
    behaviors that confer evolutionary advantage,
        77–78
    biblical creation and, 30
    and biological diversity, 3, 81
    of the brain, 54, 55
    of DNA, 3
    common ancestors in, 83–88
    environmental changes and, 83
    and ethics, 75–80
    as a generator of progress, 76–77
    genome maps for study of, 11
    of Homo sapiens, 89
    inside cells, 98–99
    of microorganisms, 82–83
    molecular studies of, 82–83
    mutations and, 82, 83–84, 101
    natural selection in, 75, 81–82, 83, 84, 101
    origins of life, 97
    oxygenic atmosphere and, 96–98

periods of life, 89–93
Precambrian, 90–93, 98
of proteins, 83–89; *see also* Molecular clocks
see also Fossils and fossil records
Extinctions
due to habitat destruction, 100, 103
ethical beliefs and attitudes toward, 100–101
and fossil record, 94
loss of genetic information through, 101–102
mass, 94, 100
oxygenic atmosphere and, 96

F

Familial hypercholesterolemia, 8
Fitch, Walter, 85
Forensic medicine, DNA probes in, 17–18
Fossils and fossil records
extinction and, 94
gaps in, 97
molecular clocks compared to, 86–87
from Phanerozoic period, 89–90
Precambrian, 90–93
of soft-bodied multicellular organisms, 89
stromatolites, 90–94
Fruit fly (*Drosophila melanogaster*), 11, 35–38

G

Gaucher disease, 13
Gene therapy
criteria for, 23
ethical considerations in, 7, 23
moral considerations in, 30
problems with, 27
prospects for, 23–27
techniques, 24–27
Genes
defined, 6
distance between, 12
dominant, 8, 11
expression of, 6, 11, 33–34, 38
number of, in different organisms, 101
recessive, 8, 13
synthetic, 24
Genetic bases, 5, 6–7, 15, 21
Genetic code, 3, 8, 15
Genetic diseases and disorders
carriers of, 8, 17
diagnosis of, 10, 17–18; *see also* Genetic testing
dominant, 8, 18
environmental interactions in, 8–9
mapping of, 10–11, 13–14, 18
monogenic, 8, 10, 13, 17, 18, 23
multigenic, 8–10, 13, 17

number identified, 10
prevalence of, 8
recessive, 8, 13
therapies for, 23; *see also* Gene therapy
Genetic linkage maps, 16
Genetic markers, 14, 15, 17, 18, 20
Genetic testing
confidentiality in, 19
ethical considerations in, 6–7, 17, 18
misinterpretation of, 19
patient understanding of, 19
social issues in, 19
use of DNA probes in, 17–18
Genetics
contributions of molecular biology to, 2–3
and disease, 7–10; *see also* Genetic diseases and disorders
early research techniques, 5
Genome
defined, 5
noncoding portions of, 82
Genome maps and mapping
fruit fly (*Drosophila melanogaster*), 11, 35
human, 11, 19
methodology, 11, 13–14
public education on, 31
rationale for, 29
recombinant DNA techniques and, 14–16
uses of, 10–11, 23, 82
see also DNA sequencing
Greenhouse effect, 98–99, 100
Growth factors
cell–cell communication through, 41
fibroblast, 41–43
human growth hormone, 31
transforming (TGF-beta), 41–43, 46

H

Heart diseases, 8
Hodgkin's disease, 104
Hood, Leroy, 6, 17, 20–23
Human growth hormone, 31
Huntington's disease, 18
Hypertension, 8

I

In vitro fertilization, 49
Infectious diseases, genetic basis for, 10
Inheritance
of altruism, 77
elements of, 2, 8
mechanisms in, 11–12